# SAFETY MANAGEMENT BEYOND ISO 45001

ANDREW YEW

PARTRIDGE

Copyright © 2019 by ANDREW YEW.

Library of Congress Control Number: 2019917247
ISBN:     Softcover      978-1-5437-5525-1
          eBook          978-1-5437-5526-8

All rights reserved. No part of this book may be used or reproduced by any means, graphic, electronic, or mechanical, including photocopying, recording, taping or by any information storage retrieval system without the written permission of the author except in the case of brief quotations embodied in critical articles and reviews.

Because of the dynamic nature of the Internet, any web addresses or links contained in this book may have changed since publication and may no longer be valid. The views expressed in this work are solely those of the author and do not necessarily reflect the views of the publisher, and the publisher hereby disclaims any responsibility for them.

Print information available on the last page.

To order additional copies of this book, contact
Toll Free 800 101 2657 (Singapore)
Toll Free 1 800 81 7340 (Malaysia)
orders.singapore@partridgepublishing.com

www.partridgepublishing.com/singapore

# PREFACE

The desire to write this book started fifteen years ago, after I had already spent almost a decade working as a safety auditor for various systems, including the International Safety Rating System (ISRS). It was a sobering experience to see firsthand that the majority of companies in reality have ineffective safety management systems. Although many world-class companies do excel at safety, it is much more commonplace to see companies that are mediocre or simply perfunctory in safety management.

The information presented during a safety audit is usually filtered to make the company look good, and only in the casual, informal talks after audits are real problems and issues openly discussed (if trust had been established). As I started to chalk up more visits to companies of various sizes and across industries, some behavioural patterns became evident.

I began to wonder why certain companies do not care for safety, and it certainly bothered me that many preventable accidents keep recurring. The technical reasons and causes for accidents have been discussed in countless books, so here I will instead delve into the less technical but crucial reasons for poor safety management.

Through more than two decades spent on the frontlines of safety auditing and consulting for numerous companies, I have occasionally chanced upon those that have found ways to effectively implement their safety management systems, sometimes despite the odds. These companies inspire me and show me that it is possible to overcome the challenges to light a way forward with safety that is well managed and truly makes a difference to human lives.

When working as a certification auditor, it quickly became obvious to me that many companies get certified simply because they need the certificate, rather than having any real goal to improve their safety. Safety management standards are often so loosely established that it is possible for a company to do very little in order to get certified, with the bulk of that work done by an external consultant with a superficial understanding of the real safety risks faced by the company.

The second part of this book thus provides guidance to companies that desire to implement a more robust system beyond the requirements of ISO 45001, such that real improvements may be made meaningfully to the safety management system.

# CONTENTS

Preface ................................................................................... v

Chapter 1    Competitive Advantage of Safety ........................................ 1

Chapter 2    Problems in Safety ............................................................ 3

Chapter 3    Inappropriate Management Response and Inadequate Safety Management ........................................ 6

Chapter 4    Deeper Causes of Accidents and Other Dysfunctions ..... 10

Chapter 5    Efficiency and Effectiveness of Safety Management System ....................................................... 59

Chapter 6    Appropriate Management Responses ............................ 72

Chapter 7    Loss Control Matrix ....................................................... 102

Chapter 8    Elements of the Loss Control Matrix ............................ 112

# CHAPTER 1

## COMPETITIVE ADVANTAGE OF SAFETY

The first duty of an organisation is to survive and the guiding principle of business economics is not the maximisation of profit, it is the avoidance of loss.

—Peter Drucker

Safe companies may not be successful, but a successful company must absolutely be a safe company. Safety is a success factor. If safety is not managed properly and efficiently, safe companies may still not attain success, because they could simply be incurring high costs without the commensurate safety outcomes. The reason is that many safety management systems are inefficient and ineffective. A strong and efficient safety management system can give a company a tremendous competitive advantage, from loss prevention to reputation and customer preference.

Companies with strong safety would generally also have a strong safety culture, which is essentially the most efficient way to prevent accidents.

The most competitive companies in any industry are the ones that do not suffer continual major losses through accidents and have the ability to prevent major accidents from happening. Accidents result in direct and indirect losses to the company. If a company is able to prevent accidents and losses, it will certainly be more competitive, provided the company spends proportionate amounts on safety.

A competitive company must also be able to consistently deliver its products and services smoothly and without mishaps or interruptions. Customers must be assured that the company is able to fulfil its promises and contracts. The company must have strong operational control over its processes, and safety is only one aspect of this. If a company cannot provide its clients a measure of assurance that production processes will go as planned, customers will not have any confidence in this company.

A successful company would have assurance that its operations do proceed as planned, with minimal deviation or incident, and that expected outcomes in terms of schedule, quality, and quantity are guaranteed. Its customers would feel confident in the company's ability to deliver on its promises, regardless of any incident.

If a company cannot ensure that employees can get home safely each day, good and talented employees will not join or stay long. Unable to attract or retain good employees, a company will eventually succumb to its competition. The best companies take better care of their employees.

These are the three main reasons why any company that aspires to be an industry leader must implement the most efficient and effective safety management system possible.

- loss control

- business strategy

- people

Finally, major accidents can cause catastrophic losses and threaten the very survival of a company, even if it is properly insured. Major accidents can lead to a crisis, with the loss of trust from regulators and customers.

Most safer companies, however, tend to be less cost-competitive than more mediocre companies. Many safety management systems are inefficient and ineffective, and they incur a higher cost. Often the easiest way to improve safety is to make operations slower, more arduous, and less efficient. This need not be the case, and the challenge is to maintain overall efficiency while improving safety.

# CHAPTER 2

## PROBLEMS IN SAFETY

> The principles governing the behaviour of systems are not widely understood.
>
> —Jay Wright Forrester

Why do we read about industry accidents almost every day in the news around the world? Why do many companies preach that "all accidents are preventable", yet they continue to suffer serious accidents every year? Why do the same accidents recur time after time? Why do complex, well-defended systems fail? Why are there so many chronic safety problems?

If the objective of safety management is to prevent accidents, property damage, and injuries, then why do we constantly have to exhort senior management to be committed to safety? Why do we make managers and workers pledge to work safely? Why do we need to write laws and contracts to force companies to work safely? Is it not common sense for a company to prevent accidents and losses?

Accidents will occur naturally and inevitably unless there is deliberate intervention. All accidents are caused, and they can be prevented by eliminating some of the causes. These causes need to be continually managed in order to prevent accidents from occurring. If the root causes of accidents can be removed, then the accident may be prevented for good. However, not all root causes may be easily removed—especially, in the chase for profits. As a result, accidents continue to happen. Still, it is possible to look for and to mitigate deeper causes so that accidents may be prevented more effectively, or for a longer duration, until other causative conditions arise again.

We frequently see companies engaging in continuous efforts to improve safety, but they lapse into unsafe practices the moment there is a shift in attention. Safety effort appears to be Sisyphean, with no long-term improvement in safety. Whole industries may even work unsafely and in a poorly regulated or poorly managed environment.

There is an urgent need to understand the deeper causes of accidents so that safety management can see real improvements. Often these deeper causes may not be totally eliminated, but they can be mitigated or ameliorated to a certain extent.

At this point of time, a high level of safety is affordable only to a minority of organisations whose value of their products and services are so high that it can justify the defences to prevent disruptions to their processes. The challenge now is to ensure that organisations who cannot afford expensive defences are able to implement cost-effective measures, and that those who can actually do so in practice.

Safety management operates in a highly complex system, where deeper causes and effects are not closely related in time or space. Cause and effect that are closely related in time and space are mainly true only in simple systems. In more realistic complex systems, causes may be far removed in both timing and proximity from their observed effects. Decisions made in the manager's office can translate into an accident ten years down the road. But investigations carried out after the accident may not be able to link it to the earlier cause, especially if there are no records. It is always easier to find the faults at the sharp end. Yet to really solve these safety problems, these relationships, however tenuous, need to be uncovered.

There are many interconnecting feedback loops in complex systems—that is, a new programme intended to solve a problem can result in reactions in other parts of the system that counteract the new programme. Success is more assured only if we are aware of the dangers.

Safety is not just about the "engineering control" of hazards but also the science of managing all the different people and parties involved in the management of safety, including their actions, behaviours, and motivations.

# CHAPTER 3

# INAPPROPRIATE MANAGEMENT RESPONSE AND INADEQUATE SAFETY MANAGEMENT

Those who cannot remember the past are condemned to repeat it.

—George Santayana

Accidents occur when there is inadequate management intervention. Inadequate management intervention is due to either wrong management response or an inadequate safety management system.

Safety management is often complicated by the fact that the decision maker may not be the person who is exposed to the risk or the one who needs to comply with the requirements.

Typically, accident investigation will identify these causes:

1. inadequate compliance
2. physical incapacity
3. mental incapacity
4. lack of knowledge
5. lack of skill

6. inadequate training

7. inadequate procedure

8. inadequate communication

9. inadequate checks and inspection

10. psychological stress

11. inadequate leadership

12. inadequate supervision

13. inadequate engineering

14. inadequate risk assessment

15. inadequate protection

16. inadequate machine guarding

17. inadequate warning system

18. inadequate purchasing

19. inadequate maintenance

20. inadequate tools or equipment

21. inadequate materials

22. inadequate work standards

23. wear and tear

24. abuse or misuse

25. poor housekeeping

These causes may be a result of deeper causes that can be classified into the following categories.

1. performance targets and profits maximisation
2. customer
3. competition
4. requirements
5. implementation efficiency and effectiveness
6. human
7. management
8. organisation
9. industry
10. complexity and nature of accidents
11. natural state of ignorance
12. constraints and pressures
13. circumvention and deviation
14. external factors

Multiple parties and stakeholders may be present in a risk situation. The outcome of a safety situation, good or bad, is the result of the combination of behaviours and attitudes of various interested parties in the system.

1. management (top, middle, and front line)
2. corporate

3. company shareholders
4. customers
5. authorities
6. public (neighbours and nongovernmental organisations)
7. industry bodies
8. competitors
9. professionals (consultants, architects, engineers, etc.)
10. auditors (first, second, and third party)
11. workers
12. contractor's management
13. contractor's workers
14. subcontractors
15. service providers, vendors, and suppliers
16. temporary workers
17. visitors

Each of these interested parties has a different interest and concern about safety requirements and outcomes.

Inappropriate safety management usually perpetuates or aggravates the risk in a system.

# CHAPTER 4

# DEEPER CAUSES OF ACCIDENTS AND OTHER DYSFUNCTIONS

> When a big vision meets a dysfunctional system, the dysfunctional system wins every time. Fix the system and success will flourish!
>
> — Daren Martin

In a majority of organisations, safety or occupational health and safety (OH&S) management is often inadequate or management's response to safety issues may be inappropriate. Safety efforts can face huge resistance, and workers' behaviours may be stubbornly unsafe the moment they are unsupervised.

A list of factors may be present in certain organisations subject to various circumstances and conditions. Typically, an organisation may be afflicted with just a few of these conditions or factors. Frequently, these factors that manifest themselves may be related to one another. Management should determine whether these factors do exist in their companies, and they must find ways to mitigate them wherever possible. Many underlying factors influence the decision-making process of a manager.

It is important to understand these deeper causes because they can manifest themselves as deficiencies and failures in different parts of the safety management system. Although a root cause analysis may identify those system failures, it does not get closer to solving the problem in the long run.

Other than slight differences, some of the factors listed below could also be related to one another. Oftentimes, a cluster of co-related factors may be present interactively, and hence some of these factors will overlap each other. However, the degree of their interrelationship may vary from one organisation to another.

The existence of these factors is dependent on a unique set of circumstances, and thus they will vary and may not be found in another organisation. It is also possible for some organisations to recognise some of the common factors and subsequently taking actions to mitigate them.

Any one of these factors, if critical enough, can undermine the entire safety or OH&S management system. Factors that can cause a safety management system to be suboptimal are listed and discussed here.

1 Performance Targets and Profits Maximisation

    1A. No Profits for Safety

    1B. Profits and Profit Maximisation

    1C. Conflict of Objectives

    1D. Cost of Safety

    1E. Cost versus Benefits

2 Customer

    2A No Desire for Safety

    2B Cheaper and Faster

    2C Not Paying for Safety

    2D Distractions

    2E Fear of Involvement

3 Competition

    3A Lowest Quote

    3B Unfair Competition

    3C Ultracompetition

    3D Bidding Mistakes

4 Requirements

    4A Lack of External Requirements, Legal or Other

    4B Inadequate Checks and Enforcement

    4C Certification and Superficial Compliance

    4D Inadequate Internal Requirements and Standards

    4E Unrealistic Requirements

    4F Superficial Requirements

    4G Ambiguous Requirements

    4H Hollow Buzzwords

5 Implementation Efficiency and Effectiveness

    5A Paper Exercises and Going through the Motion

    5B Disconnected and Poorly Designed Management Systems

5C Ineffective/Inefficient Control Measures and Poor Fixes

5D Progressive Reduction in Efficiency

5E Depleting Resources on Noncritical Hazards

5F "Flavour of the Month" Safety

5G Negative Process

5H Lack of Compliance

5I Ultrasafety

5J False Indications

6 Human

    6A Errors, Slips, Lapses, and Mistakes

    6B Laziness, Easier Ways, Rushing, and Shortcuts

    6C Worker Risk Attitudes and Complacency

    6D Safety Creep

    6E Resistance to Change

    6F Habits and Practices

    6G Overload, Confusion, Distractions, and Fatigue

    6H Foreign Cultural Differences

    6I Improper Motivation and Production Incentives

    6J Fatalism

7 Management

    7A Inadequate Control and Management

    7B Lack of Safety Management Experience

    7C Inadequate Safety Function

    7D Avoiding Self-Blame and Aversion to Criticism and Deficiencies

    7E Supervisory Conflict

    7F Friends and Colleagues

    7G Risk Perception

    7H Counterintuitive Effects

    7I Safety a Staff Function Responsibility

    7J Focus on Lagging Indicators and Equate "No Accident" as Safety

8 Organisation

    8A High-Risk Appetite and Risk-Taking Culture

    8B Poor Safety Culture

    8C Power Dysfunction and Safety Leadership

    8D Lack of Line Ownership

    8E Risk Habituation and Normalisation of Risk

    8F Business Risk and Risk Aversion

    8G Dogma and Political Incorrectness

8H Cyclical Commitment

8I Poor Safety Organisational Structure

8J Lack of Talent

9 Industry

9A Lack of Industry Standard

9B Inadequate Sharing of Good Practices

9C Inadequate Sharing of Incident Learning

9D Inappropriate Industry Standards for Safety

9E Criteria for Safety Awards

9F Safety as a Market Entry Barrier

9G External Funding and Subsidies

9H Subcontractors and Sub-Subcontractors

10 Complexity and Nature of Accidents

10A Technology

10B Infinite Possibilities

10C Probability, Chance, and Unpredictability

10D Safety Outcomes Not Proportional to Safety Effort

10E Murphy's Law

10F Flawed Investigation and Root Cause Analysis

10G Knee Jerks after Accidents

11 Natural State of Ignorance

    11A Workers

    11B Management

    11C Customers

    11D Inadvertent Deviations

12 Constraints and Pressures

    12A Lack of Resources

    12B Personal versus Organisational Resources

    12C Inadequate Time

    12D Delays and Liquidated Damages

    12E Poor Planning and Scheduling

13 Circumvention and Dysfunction

    13A Corruption of Function

    13B Voluntary and Involuntary Scapegoats

    13C Blame Externalisation

    13D Fraud and Scams

    13E Abuse of Safety Argument

13F Office Politics

   13G Suppressed Incident Reporting

   13H Paid External Audits

   13I Forgery and Paperwork

   13J Defensive Safety Management

14 External Factors

   14A Uncontrollable Factors

   14B Unknown Causes

   14C Business Environment

   14D External Safety Function

1 Performance Targets and Profits Maximisation

The primary purpose and goals of an organisation will usually be prioritised over safety considerations because they are the raison d'être for the existence of the organisation. These primary goals and safety goals often come into conflict.

   1A No Profits for Safety

   By itself, safety activities do not make any profits for the organisation. Worse, if costly and unnecessary safety provisions are implemented, profits will be reduced.

   Because safety is mostly a nonprofit activity, there is a tendency for its activities and resources to be reduced rather than increased.

## 1B Profits and Profit Maximisation

Companies may undertake jobs that they cannot do safely, taking intolerable risks for the sake of chasing profits. Companies may also fail to allocate money and resources to reduce risks so as to maximise profits in the near term.

## 1C Conflict of Objectives

The first objective of any business is to make profit, and safety issues may well come in the way of that, resulting in a conflict. In many workplaces, there are often conflicts between safety and the profit motive. Managers are typically required to ensure that production activities are not interrupted or slowed. For example, should production activities stop if they are found to be unsafe? Should production be carried out in a faster but less safe manner?

Organisations exist for a purpose or mission and this purpose or mission will often override the need for safety, unless management is capable of managing the safety risk. If management feels that there are inadequate resources to achieve the organisation's objectives safely, they will often find other strategies to hide or deflect the safety risk from themselves.

There is often an urgent need to get activities going and make progress, before necessary or adequate safety considerations and provisions have been made. Safety risks are often taken to achieve the overall objectives of the organisation. For example, production processes are not slowed down or stopped even if there is a safety concern. A faster way of production or shortcuts may be taken instead of a slower and safer process.

All parties may push for progress if there is a potential delay in a project's timely completion. Unforeseen delays to projects may exacerbate safety problems.

Safety may be neglected when other work-related tasks are given higher priority than safety tasks.

## 1D Cost of Safety

The total cost of all safety measures and provisions has to be paid for, usually from the profits made, or by the customers. If profits are insufficient to pay for these measures, then safety measures would not be sustainably maintained if they have not been priced in.

An organisation may choose to do without proper equipment, training, or competence levels in order to save cost and increase profits.

Safety costs money, time, and effort in the following areas:

- hiring qualified and competent personnel
- providing training and awareness
- providing proper equipment
- providing safety equipment
- doing safety work, risk assessments, safety checks, and inspections

There is also a tendency for vendors and suppliers to charge a very high price for their safety products and services if they are seen as a premium and a requirement, and if there is insufficient competition.

Organisations may attempt to reduce the cost of safety in counterproductive ways:

- appointing less competent or less safety personnel as needed

- using substandard equipment and materials

### 1F Cost vs. Benefits

In 1A, it has already been said that safety activities per se do not make any profits. Nevertheless, when companies put X dollars into safety, there is an expected return of Y dollars in the form of the reduction in expected losses. To make sense, the benefits (Y) should be higher than the costs (X), but frequently the cost of a safety effort may exceed the benefits received. From a financial point of view, it does not make sense to implement such initiatives unless there are legal or other requirements to do so. Typically, companies will try to avoid implementing such measures. Such low-benefit initiatives are not sustainable, and if sustained, they are usually paper exercises or are not properly implemented.

## 2 Customer

Customers are the raison d'être for companies. Customers' interest in safety, or the lack of, has a direct influence of whether a company will properly manage safety.

### 2A No Desire for Safety

Frequently, customers are only interested in the final product and deliverables; they are not really interested in whether or not the product is made in a safe manner. Occasionally, accidents may impact the customer in the form of delayed delivery or damage to customer's property. In any case, customers may not care much about whether work is carried out safely in the making of the product and thus will not be willing to pay any premium for safety.

## 2B Cheaper and Faster

Customers are mostly concerned with products and/or services being delivered as quickly and as cheaply as possible. They may pressure the manufacturer to produce the products more cheaply or quickly, which may have negative impacts on safety (e.g., property developers may rush a contractor to complete a project on an unrealistically short project schedule).

## 2C Not Paying for Safety

Customers will pay only for the product or the service that is rendered, frequently not caring whether the contractor or service provider is taking risks to produce the product or service. Consequently, customers will not pay extra for any safety measures taken to protect workers. They are often not concerned with how the product is made or service provided unless it affects them directly. Thus, they usually will not pay for safety of the manufacturer or service provider.

## 2D Distractions

Some customers may be interested in safety, but they get involved in unproductive ways. They may request companies do noncritical safety errands that deplete safety resources (e.g., getting the safety manager to attend various nonessential meetings, or getting safety personnel to investigate and report on trivial incidents, organise inconsequential activities).

## 2E Fear of Involvement

External parties, including customers and their representatives, may fear directly involving themselves in safety processes because in the event that something goes wrong, they may get a share of the blame or legal liability. This is despite knowing that their involvement can contribute to safety and reduce risk.

## 3 Competition

The survival of a company depends on whether it can effectively compete. If implementing safety makes a company less competitive, the company will find ways to bypass safety.

### 3A Lowest Quote

Customers may award the job or contract only to the lowest bidder without regard for their safety performance or capability. The lowest bidder may not have catered for all the required safety provisions and may even plan to avoid providing all the necessary safety provisions.

### 3B Unfair Competition

This refers to a situation where a competitor can enjoy a cost advantage by working unsafely. Customers may not check that contractors are implementing safety requirements in their contract specifications, resulting in an uneven playing field. Contractors working unsafely can reduce their operational costs and bid a lower price, albeit with great risk-taking or noncompliance to certain safety requirements.

Customers may also knowingly condone such a situation to save money after all the risks may already been transferred to the contractor.

### 3C Ultracompetition

There are many reasons that can result in industries becoming ultracompetitive. For example, certain industries may have cycles and during low periods due to shortage of demand, or prices may become ultracompetitive. When profit margin is eroded, companies will find ways to cut costs to minimise losses. Often the first things to be cut are safety personnel and/or safety provisions.

### 3D Bidding Mistakes

Mistakes may have been made during the bidding process. The costs of certain major items or safety provisions may have been missed out, leading to a contract being awarded to an extraordinarily low bid.

In the execution of such projects, the budget for safety provisions may be severely curtailed, leading to cost-cutting and underprovision of safety measures.

## 4 Requirements

Safety requirements may be lacking because multiple forces are at play with multiple interested parties. Stronger stakeholders may not be interested enough in safety to solve real safety problems which they do not see as belonging to them. Or worse, inappropriate or unsuitable requirements may be arbitrarily imposed. Requirements, when established, often serve the needs of the external party first rather than the companies that are required to implement them.

### 4A Lack of External Requirements, Legal or Other

For various reasons ranging from ignorance to lack of interest, customers or external bodies may not have established any meaningful safety requirements. These external stakeholders may not see themselves as "owning" the set of safety problems, and solutions could also be complicated. Moreover, should their proposed solutions fail, they may receive some blame. It is frequently easier to shift the blame directly to the companies involved in the accidents.

Customers may not have included adequate contractual requirements for safety provisions in the contract documents.

Although many high-risk industries may have many requirements for a "well defended system", and failures only occur when the swiss cheese is breached, many low-profit industries may not have much requirements or even any swiss cheese at all.

### 4B Inadequate Checks and Enforcement

Parties setting the requirements do not meaningfully check or enforce the requirements. Sometimes this may be done intentionally because there is no real desire to force the other parties to comply and incur additional costs.

Weakness in internal audits and checks can often be seen because these activities by themselves do not make profits and are additional costs in terms of time, effort, and resources.

### 4C Certification and Superficial Compliance

Certification is a convenient requirement to impose on service providers, but by itself it does not guarantee that an effective safety management system will be implemented. Many organisations may have certified systems, but these systems have variable levels of effectiveness. In the worst cases, a bare-bones system may be implemented for compliance with the required standards, but there is no substance them. Yet some customers are easily satisfied by many of these certificated systems. There are even a few organisations with certified systems that have almost zero effect on accident prevention. Not only are such systems ineffective, but they give a false and potentially dangerous sense of assurance. Interviews with employees may even reveal that they are completely unaware that the company is certified or has implemented a safety management system.

These organisations may comply superficially with requirements, but the implemented systems are not effective in preventing accidents.

People may comply with requirements superficially by simply ticking off a checklist.

Although getting companies to be certified is useful as another convenient layer of check, certification by itself may not be adequate for assurance. Only the best certification bodies would be able to provide the minimum required level of assurance.

4D Inadequate Internal Requirements and Standards

Internal requirements are inadequate for work to be carried out safely. Because requirements have not been set, implementation will be inconsistent at best and will vary between departments or projects. If there are no internal requirements, enforcement is also not possible.

4E Unrealistic Requirements

> *Remedy is worse than the disease.*
>
> —Francis Bacon

When requirements are established by external parties that do not have to comply with or implement them, these requirements may be set unrealistically high. If requirements set are too high and cannot be sustained, people will find ways to circumvent these requirements.

As mentioned earlier, the total cost of safety measures has to be paid by the profits made. If the profits are insufficient to pay for these measures, then these measures will not be sustainable.

## 4F Superficial Requirements

Safety requirements established may be superficial simply to show stakeholders that some risk management is in place or something is being done. Superficial requirements are typically to placate other ignorant stakeholders. Sometimes it is because the people in charge may not have a full understanding on what needs to be done to reduce risk, resulting in requirements that may have no real effect in reducing OSH risk.

## 4G Ambiguous Requirements

When legal and other requirements are ambiguous, it leads to confusion and wasted effort. Sometimes requirements are purposely made ambiguous so that it can be interpreted in the most advantageous way to the author. Sometimes clarifications may also not be given when asked, or they could be given privately or verbally and cannot be relied upon.

## 4H Hollow Buzzwords

A buzzword is a word, phrase, or jargon that is fashionable at a particular time or in a particular context. Buzzwords may be useful in conveying ideas and concepts, but if management uses these buzzwords with little intention of following through, then they become hollow and lose their meaning.

Common buzzwords are "Safety First" and "Zero Accidents".

High-sounding buzzwords are often high-level requirement that translate into little at the implementation stage.

Due to the cost or difficulty of safety provisions, a situation often described as NATO ("No action, talk only) frequently results.

Buzzwords, where fully resourced and supported by management, could be useful; otherwise, management loses credibility in the eyes of workers and is detrimental to the safety culture.

## 5 Implementation Efficiency and Effectiveness

Good systems may not deliver good results if they are not properly implemented. Systems and procedures often look good on paper but deliver dismal results in real life.

### 5A Paper Exercises and Going through the Motion

Paper exercises are useful for

- providing a false sense of control and confidence,
- shifting the blame,
- pinpointing scapegoats, and
- furnishing evidence for defence.

Paper exercises are generally ineffective in preventing accidents but often serve as a convenient "solution" to certain types of intractable accidents.

Frequently, it can be observed that systems are superficially implemented with minimum effort and no real intention to ensure effectiveness in risk reduction. Usually this is due to a belief that the required activity does not really benefit the person in charge or the organisation.

### 5B Disconnected and Poorly Designed Management Systems

Many management systems are poorly designed and often follow a generic template that may not be suitable or efficient for the organisation. Processes and systems may not be customised to fit the organisation to provide effectiveness and efficiency. Many management systems are simply "off the shelf" and "cut and paste" jobs done by external consultants.

Such systems do not address the real operational risk faced by the companies.

5C Ineffective/Inefficient Control Measures and Poor Fixes

A high-cost and inefficient system is not sustainable because the motivation to maintain such a system will erode day by day. An ineffective system will last only as long as status quo can be tolerated. It does not give the needed payback or produce the required results.

More often than not, control measures are poorly designed, or the person setting the standard is different from the persons needing to comply. Poor control measures often lead to noncompliance, and this may lead to accidents. Inefficient work processes frustrate workers, and they may find alternative ways and shortcuts to complete the job. A very common way to "increase" safety is to make workers do the job slower or in a more circuitous way.

5D Progressive Reduction in Efficiency

Many implemented systems will experience a reduction in efficiency as more measures are added and they go beyond the point of diminishing returns. Usually, the newer measures are more difficult to implement and take up more resources. They may overlap with existing practices as well. Thus, although the overall level of safety may still increase, the efficiency of the system usually falls because much more work is done to achieve only marginally better performance.

The Safety Management System may also be hijacked to take on other management programmes that may or may not be related to safety, thus distracting its focus and consuming available resources.

## 5E Depleting Resources on Noncritical Hazards

Many companies may rush into safety seeking to make the environment only as safe as they know how to with limited expertise. They may be content because they seem to be doing many things and burning up resources. Or a company may not be able to properly prioritise and identify critical safety issues that need to be managed. Simply "doing safety" is not enough because it is critical to do the important things first. For example, many companies focus on asking their employees to hold on to handrails when climbing stairs, whereas major hazards in the workplace such as fire and explosion have not been controlled simply because this is the easier thing to do.

## 5F "Flavour of the Month" Safety

Many safety management systems and programmes are not based on the real needs and risk exposures of the organisations; instead, they are based on fads and "flavours of the month".

The organisations lurch from one safety programme to another without many results to show for them because they are not based on risk or needs. Such fads may also last for years instead of months.

## 5G Negative Process

Safety management is often a negative process because it is always identifying

- risks and hazards,
- unsafe behaviours and unsafe conditions,
- injuries and losses,
- lapses and mistakes,
- nonconformance and violations,
- warnings and reprimands,
- punishment and discipline,
- fault-finding and accountability,
- inspection findings and audit findings,
- immediate causes and root causes of accident, and
- prosecution and fines.

This can be depressing and demotivating over time. If the safety function is largely negative, it will ultimately become dysfunctional and degenerative.

Safety may also be perceived as a negative process by the organisation.

In some organisations, safety personnel have become "policemen", with safety degenerating into a cat-and-mouse game. With "safety enforcers", relationships with colleagues in other departments may be strained. Safety personnel often also have to constantly nag workers to work safely.

In any case, "safety" is often one of the nonglamorous parts of management function because it does not generate any profit, no credit is given when things go right, and one is blamed when things go wrong.

## 5H Lack of Compliance

Because there is cost and effort to comply with safety requirements, noncompliance is a possibility if there are no checks and consequences. Management itself may simply choose not to comply with the requirements unless regularly audited. Workers will choose not to comply with requirements if they are inconvenient or uncomfortable and there is inadequate enforcement.

## 5I Ultrasafety

In some organisations, safety is seen as all good and no bad, and all available resources are put into safety. Is a ten times safety factor good? Is there a need to reduce the probability to $10^{-6}$ when the risk can be tolerated at $10^{-3}$? Is a $1 million engineering control better than a $2 PPE (Personal Protective Equipment) if that can fix the problem despite the "hierarchy of control"? Going beyond what is necessary is very often a waste of resources. A safety factor of ten simply means 70 to 80 per cent of the resources allocated may be wasted. These resources can be more meaningfully allocated to other areas of safety.

## 5J False Indications

A "false indication" situation arises when positive results are shown for safety indicators but may not be totally relevant.

Examples:

- Excellent housekeeping does not equate to excellent safety.

- Excellent "safety culture" does not equate to excellent "safety system".

- Excellent "personal safety" does not equate to excellent "process safety".

- "Zero noncompliance" found does not equate to full or meaningful compliance.

Particular aspects of a safety management system might be performing very well, and management may interpret this as an excellent state of safety. This may lull management into a false sense of security and to conclude that risks are well under control thus making no significant changes or improvements.

## 6 Human

> Workplaces and organisations are easier to manage than the minds of individual workers.
>
> —James Reason

Humans are probably the single most difficult factor to manage in any accident causation chain. Nevertheless, conditions influence human performance and behaviours, and it is still the responsibility of management to manage these conditions.

### 6A Errors, Slips, Lapses, Mistakes

Human errors, slips, and lapses may be made repeatedly, leading to accidents. Some processes and designs are more prone to human errors and lapses.

### 6B Laziness, Easier Ways, Rushing, and Shortcuts

Some workers may be lazy and may not strictly follow procedures. Workers may choose an easier but unsafe way or take shortcuts to complete the job. Workers may rush to complete a job so that they can be free to do other things. Workers may bypass interlocks and disable alarms that are inconveniencing, delaying, or irritating.

6C Worker Risk Attitudes and Complacency

Workers may have different risk attitudes, and some will be more inclined to take risks. Risk attitudes are affected by factors such as the following.

- overestimating capability or experience
- familiarity with task
- seriousness of the outcome
- being in control
- personal experience
- cost of noncompliance
- confidence in equipment
- confidence in protection and rescue
- potential profit or gain
- role models accepting risk

Ironically, workers may be more likely to take risks if they think that safety measures are in place.

When there has not been any incident for a period of time, workers may assume that an accident will not occur as long as they do things in the same old way. Workers and supervisors may be less vigilant, take things for granted, or make certain assumptions that inure or predispose them to take risk.

## 6E Resistance to Change

> If it ain't broke, don't fix it.
>
> —Thomas Bertram Lance

Management and workers may adopt a "If it ain't broke, don't fix it" attitude. If they have been working unsafely for the last ten years but no accident has occurred, they may be reluctant to make any changes or improvements.

## 6D Safety Creep

Over time, there may be an unintended relaxation, or creep, in safety practices. Riskier practices may begin to be tolerated if there is no vigilance against such development. Minor steps taken for safety may be forgotten or overlooked. Creep may be unconscious, and there may not be any deliberate attempt to undermine safety.

## 6F Habit and Practices

Once a particular behaviour has become a habit or practice, it will require tenfold effort to change this habit. Unsafe habits and practices may never be completely eradicated, even with constant monitoring and enforcement.

## 6G Overload, Confusion, Distractions, and Fatigue

If people are overloaded with information, they may get confused, and the likelihood of mistakes is exacerbated.

People can also easily lose their concentration if there are distractions nearby or if they are tired.

## 6H Foreign Cultural Differences

Safety solutions (e.g., behaviour-based safety, safety culture initiatives) developed in one country may be transplanted wholesale into another country or culture without much cultural consideration or modification. What is fine in one culture may be taboo in another culture. Under different circumstances, some initiatives may have the opposite effect.

## 6I Improper Motivation and Production Incentives

Management often tries to motivate production performance by tying incentives to production targets. Workers may be improperly motivated to rush production activities and take risks.

Safe behaviours may be disincentivised or punished for being slower.

## 6J Fatalism

In certain places, people may have a greater belief in fatalism. Fatalism is a philosophical doctrine that stresses the subjugation of all events or actions to destiny. People may have the view that they have no power to influence the future or prevent any accident and that accidents, deaths, and injuries are not preventable. Fatalism may be further exacerbated by related religious beliefs.

Fatalists may take certain risks because they feel that the outcome has already been predetermined.

# 7 Management

Most of the root causes of accidents can be attributed to management and management system deficiencies. This is a major area for improvement.

## 7A Inadequate Control and Management

Inadequate safety management is another result of wrong safety management and is typically cited as a reason why accidents happen. Management may not exercise sufficient control on operations for a multitude of reasons ranging from ignorance to incompetence to lack of resources. Inadequate control and management may be due to inadequate system or programmes, inadequate standards or requirements, or inadequate compliance at various levels.

## 7B Lack of Safety Management Experience

Most managers are business leaders, and they typically will not have direct experience with safety management. Management may be unfamiliar with the peculiarities and pitfalls of safety management and commit mistakes despite their good intentions. Managers may apply the wrong safety management concepts or even divert the safety function from effectively doing its work.

It is not uncommon to see top managers reacting negatively and punitively to incident and near-miss reports and not recognising it as a healthy sign of effective safety management.

## 7C Inadequate Safety Function

Safety function may be inadequate in quality or quantity. Because the safety function does not bring in revenue or profit to the organisation, it is frequently under-resourced. Safety staff hired may not be adequately competent or experienced. They may not have sufficient safety knowledge or experience to

- solve safety problems,

- provide safety advice and recommendations,

- investigate incidents,

- develop a comprehensive safety management system,

- effectively implement the system,

- introduce industry best or good practices, or

- address legal and other requirements.

Some organisations may be reluctant to further develop the qualifications or competence of their safety personnel because this might make them more attractive to other companies, or they may ask for higher remuneration.

7D Avoiding Self-Blame and Aversion to Criticism and Deficiencies

We have met the enemy, and he is us.

—Walt Kelly

In most accident root cause analyses, the fault will usually lie with management because management has control of resources, work activities, and risk-control measures. Most analyses will point the finger at management, but management may try to deflect this blame.

In the aftermath of an accident, if numerous corrective actions can be taken by management, then questions may also be asked as to why these actions have not been taken earlier. Is management at fault for not implementing safety measures?

In order to avoid surfacing these tough questions, management may therefore not properly address the real issues or the required control measures suggested and implemented.

Some management can respond poorly to criticism, comments, suggestions, and negative audit findings. They are more focused on defending their actions than considering any suggestions for improvement.

## 7E Supervisory Conflict

Supervisory personnel often need to maintain a cordial relationship with the workers to achieve performance targets. Thus, to avoid conflict, they may choose to "close one eye" if there are any safety violations by workers and not discipline them. Supervisory staff may then rely on safety enforcement personnel to enforce and mete out punishments.

## 7F Friends and Colleagues

People form relationships and become familiar with each other when they work together. When safety personnel have a strong relationship with other employees, it can be a strength when safety initiatives need to be implemented. But if these relationships prevent safety personnel from doing their jobs, it can become a problem. Safety personnel may look the other way when there is a violation by colleagues and friends. Or it may become difficult for safety personnel to enforce requirements on their colleagues, especially if there is a possibility of retaliation or threat to withhold future cooperation.

Friendships may also be formed with contractors, subcontractors, vendors, and suppliers, potentially impeding safety personnel from impartially performing their job.

Workers may also be reluctant to report near misses if it involved their friends and colleagues because they fear it may get them into trouble.

## 7G Risk Perception

Risk perception is the subjective judgement that people make about the characteristics and severity of a risk. The cognitive dimension of risk perception relates to how much people know about and understand risks. The emotional dimension relates to how people feel about risks. Risk perception may be irrational, and management may not deal with risk in the most logical manner to ensure critical risks are adequately controlled.

## 7H Counterintuitive Effects

Sometimes management personnel have good intentions, but their actions for safety may have counterintuitive effects. Here are some examples.

- harsh punishment for minor safety violations
- punishing workers unfairly for being involved in accidents
- finding and punishing workers' faults only during accident investigations
- making "zero accident" a requirement
- rewarding "no accidents"
- responding negatively to reporting of incidents and near misses
- forcing workers to report near-misses

- extending accountability into blaming and fault allocation

Although a "zero accident" goal is good, people should not be counterproductively punished for reporting accidents and incidents.

7I Safety a Staff Function Responsibility

In some organisations, safety is solely the responsibility of the safety staff function. Often this results in a "police and thief" situation, with safety personnel catching workers working unsafely. Worse, the safety function may be normally dormant and only required to work when accidents happen or to take the blame for any accidents.

The safety staff function often cannot decide how work is to be carried out, so any safety improvement is difficult to implement. Or the safety function may implement measures without any regard for ease and efficiency, resulting in an acrimonious relationship with other departments.

In some organisation, the safety function is more focused on justifying its existence and may not work synergistically with the line departments.

7J Focus on Lagging Indicators and Equate "No Accident" as Safety

Organisations often like to give rewards for no accidents. When there is no accident, it could simply be because of the element of luck or under-reporting rather than safety efforts being effective.

Sometimes an organisation without any safety management system may attain zero accidents, whereas another organisation with an excellent safety management system may experience bad accidents despite their best efforts.

Requiring "no accident" frequently results in accidents being hidden or not reported instead.

If "no accident" is considered as "good safety" and rewarded accordingly, resources for safety measures may be reduced during "accident-free" periods instead.

## 8 Organisation

Each organisation has its own set of values, beliefs and behaviours which may or may not be conducive to effective safety management.

### 8A High-Risk Appetite and Risk-Taking Culture

> There are two times in a man's life when he should not speculate; when he can't afford it and when he can.
>
> —Mark Twain

Management and business owners may take on high risks (e.g., take on jobs that the company may not be familiar with or competent at). Past company successes may be a result of taking extraordinary risks, and the company may continue with a risk-taking culture in the pursuit of profits.

Managers may have high-risk appetites for various reasons (e.g., faster operations, higher production, ease of operations). Managers may also be able to find ways to transfer the risk to other parties (e.g., contractors, external consultants).

### 8B Poor Safety Culture

Organisations typically start with a weak safety culture because management and workers are initially ignorant or uncaring about the importance of safety, especially if no incidents have occurred before. Poor safety culture may persist within the organisation, leading to wrong or weak management decisions and unsafe actions by workers.

Companies keen on safety may also lapse into a blame culture, which can damage a healthy safety culture.

### 8C Power Dysfunction and Safety Leadership

Related to a poor safety culture, safety leadership is usually weak unless leaders have been exposed to a strong safety culture environment and understand the importance of safety management to business success.

Senior managers often have the power to shift responsibility and accountability away from themselves. When things do go wrong, it is also possible for them to shift away some blame.

In certain companies with poor safety culture, there is an occasionally observable phenomenon where safety compliance is inversely proportional to the power of the person (e.g., lesser PPE is worn by the more senior person or VIP).

### 8D Lack of Line Ownership

In some organisations, responsibility for safety management has become mostly a staff function, and safety personnel are held accountable for all incidents.

Line departments have to carry out the core work activities of the organisation, and they may not be concerned with working safely or managing safety. Production is usually their first priority, and safety management is seen only as a peripheral activity or even as a burden.

### 8E Risk Habituation and Normalisation of Risk

Normalisation of risk is related to people's risk attitudes. Over time, people get used to the prevalent level of risk, and this increases their risk tolerance. Risk may also become normalised for the entire organisation.

## 8F Business Risk and Risk Aversion

> Learning begins when people feel safe enough to take risks.
>
> —Anonymous

Risk-taking can be a key to success for people and businesses. Risk comes with opportunity, and the fear of taking risks may mean that these business opportunities are lost. Aversion to business risk may be detrimental to the survival of the organisation in the long term. There is a difference between taking business risk and operational risk, and this distinction needs to be recognised.

To avoid blame, decision makers in an organisation may make suboptimal risk management decisions that safeguard themselves at the expense of the organisation. Unnecessarily expensive or onerous risk decisions may be made because this cost is not personally borne by the manager or department. As such, measures are inefficient and unprofitable, the organisation suffers the losses, and the systems become more inefficient over time.

In a risk-averse organisation, when safety in the organisation is further increased, the safety management system usually becomes increasingly inefficient, and resources are wrongly allocated, reducing the benefits from implementing the system.

## 8G Dogma and Political Incorrectness

Safety management can become dogmatic in organisations such that it becomes politically incorrect to suggest a more optimal alternative that may be perceived to be less safe. Safety decisions are made in a rigid way without transparency, rationale, or logic, with a tendency to continue past practices so that blame can be avoided if things went wrong instead.

### 8H Cyclical Commitment

Management personnel's commitment to safety waxes and wanes depending on whether they are recent events that triggers management attention on safety and the availability of resources.

### 8I Poor Safety Organisational Structure

Safety organisation is not properly structured if the safety function has to report to operation or production instead of directly to a higher management, because line departments may make decisions in favour of production priorities over safety. Defective safety organisation may also result in organisational influences that hamper the proper operation of the safety function. Safety function may be denied its independence to perform its role impartially or professionally.

### 8J Lack of Talent

Because the safety function is a small staff function compared to line function, there is usually less opportunity for career progression. The safety function may also encounter many frustration and barriers in performing its role. Thus, many talented, experienced or qualified personnel may be reluctant to work in the safety function.

The safety function is often required to perform a safety policing function and may need to enforce safety requirements on friends and colleagues. This may be a socially isolating position, and some people may want to avoid being placed in such a difficult position.

## 9 Industry

Each industry has its own safety culture due to the nature of the industry and the accompanying regulations and requirements (or their absence).

### 9A Lack of Industry Standard

The industry does not have adequate standards for work to be carried out safely.

### 9B Inadequate Sharing of Good Practices

Effective good practices and safety solutions are not shared in the industry. Companies may treat certain safety practices as confidential information.

### 9C Inadequate Sharing of Incident Learning

Learning from incidents is not shared in the industry, or sufficient information is not shared that can facilitate learning and prevention. Such information may not be shared for legal, reputation, political, or other business reasons.

### 9D Inappropriate Industry Standards for Safety

Different stakeholders have different interests in the setting of industry standards. Some will want lower standards, whereas others want to set higher standards. The cost of bearing the standards will not be equally borne by all the stakeholders. When standards are set too low, risks may not be adequately controlled.

When standards are set too high, they may not be sustainable, and parties will find ways to circumvent these requirements.

## 9E Criteria for Safety Awards

The primary purpose of safety awards is to motivate companies to achieve higher safety performance.

For business reasons, companies may be motivated to win safety awards. However, the criteria for safety awards may result in undesired practices.

- suppression of incident reporting
- suppression of learning from incidents
- not sharing good safety practices
- faking system implementation
- more paper exercises
- allocation of resources to noncritical safety items

This may be due to the following reasons.

- accident statistics as an award criterion
- superficial evaluation process
- focus on paperwork and paper evidence
- no assessment of actual working conditions

## 9F Safety as a Market Entry Barrier

Safety requirements may become entry barriers to a market for new companies. Requirements may be made hard to comply. Incumbents may try to raise safety standards not because there is a need but as an entry barrier to newcomers.

9G External Funding and Subsidies

Sometimes external funding and subsidies are provided to companies in certain industries to establish and implement a safety management system or programmes. Often a consultant will be hired to produce a system to meet certain requirements. In the long term, such systems are often not sustainable.

A safety management system has to deliver value from day one in order to make a business case for itself. If management cannot find reasons to justify implementing a safety management system, the temporary injection of money will not solve this problem anyway.

Such systems, transplanted by consultants, are often mere paper exercises. Paper exercises erode people's faith in the system to produce real results.

External funding and subsidies are usually given to show that something is being done to improve safety and that the work done is primarily for the purpose of claiming the money.

The performance indicator for the disbursement agency could be just the number of projects completed because it is hard to assess quality, effectiveness, and long-term benefits to the organisations.

9H Subcontractors and Sub-Subcontractors

Work awarded to contractors may actually be carried out by subcontractors or even sub-subcontractors. The workers finally carrying out the work may not even be known when the contract was awarded. Certain industries may operate in such a manner, profits are skimmed off at each level, and the final work is carried out by the workers willing to do the work at the lowest price. Sub-subcontractors may not be able to meet the contractor selection criteria at the onset. Safety culture of the sub-subcontractor may be totally different from the main contractor that has been awarded the contract. Because sub-subcontractor's workers are somewhat removed from main contractor's management and supervision, their competence, behaviour, motivation, and safety culture may be totally different from the main contractor's. Control over sub-subcontractors' workers may be ineffective or inadequate. The main contractor may not assume full responsibility for sub-subcontractor's workers.

Safety responsibilities may not be clearly assigned to subcontractors, or they could argue that it is the job of the main contractor.

If main contractor does not pay the subcontractors on time, the subcontractor may refuse to do their scope of work properly or safely.

The main contractor may not require the subcontractors to do their work safely if the subcontractors have quoted a very low price.

The main contractor's personnel may not be able to communicate properly with subcontractors' workers if they are of a different nationality.

## 10 Complexity and Nature of Accidents

### 10A Technology

Technology may increase the complexity of operations such that not all consequences may be foreseeable and controlled (e.g., MCAS plane control software, automation).

## 10B Infinite Possibilities

Many situations have infinite possibilities, and as long as all accident possibilities have not been eliminated, accidents may still occur.

Black swan events may be a possibility even if they are unforeseeable.

## 10C Probability, Chance, and Unpredictability

Many accidents are chance events, and their occurrence is not predictable. There is a strong element of chance in safety (e.g., a worker may simply step away from the line of fire for a second and avoid an accident). Accidents may also occur in the freakiest, unforeseeable manner.

## 10D Safety Outcomes Not Proportional to Safety Effort

Unlike other areas of management such as quality, safety results are often not proportional to safety effort. Operations without safety controls may not have any accidents, whereas operations with many controls may still experience accidents. Safety is unlike quality, where inputs into quality will definitely result in a better-quality product. Management may be tempted to take chances with safety or discouraged to put resources into safety if payback is not certain.

Some accidents are more easily prevented than others. Sometimes it is possible to attain "zero accident" without any effort, whereas at other times an accident may still occur despite the best efforts.

## 10E Murphy's Law

Murphy's law states, "Anything that can go wrong will go wrong." A possibility will occur eventually with sufficient time and risk exposure. An accident that *can* happen will happen sooner or later. The only way to definitively prevent a particular accident is to eliminate its possibility.

10F Flawed Investigation and Root Cause Analysis

The continued occurrence of accidents is because we are unable to eliminate all the causes of accidents. Many causes, as discussed in this chapter, are very difficult to remove. Some "root causes analysis" methodologies stop at deficiencies in the safety management system and failures of barriers, but they do not go beyond. However, there may be little efficacy in going beyond because deeper causes are often more difficult to remove. But even before root cause analysis, it is common to see the obfuscation of the real causes of an accident, especially if it is a major accident. This is mostly done to deflect blame and responsibility.

10G Knee Jerks after Accidents

Major accidents often produce knee-jerk reactions from various stakeholders for various reasons.

- to avoid blame
- to assign blame
- to show management's interest in safety

Typical knee-jerk reactions include the following.

- introducing unnecessary safety measures

- introducing unsustainable and overly risk-averse safety measures

- addressing the wrong accident causes

- finding scapegoat causes to hide the real causes

- doing something without understanding, just to show action

- doing multiple actions but cooling off quickly once the heat is off

- blaming and removing good safety personnel

## 11 Natural State of Ignorance

The initial state is a state of ignorance.

### 11A Workers

Workers typically start off with little awareness of safety, safety knowledge, and requirements. They may have little understanding of the risk and consequences.

Workers may be working for subcontractors that quoted the lowest price. Due to mismanagement, the whole spectrum of problems may be present (e.g., low safety awareness, competence and training, substandard equipment).

### 11B Management

People and organisations may start off with close to zero safety management knowledge. Safety knowledge and practices are slowly gained or learned from experience or incidents. Without any training, managers will have little knowledge on how to manage safety and provide safety leadership.

### 11C Customers

Customers are naturally ignorant about safety at the beginning because they are interested in only the final product or service. The various issues with customers have been discussed earlier. Because customers are the most powerful force for any business, they can be the key to solving many chronic safety problems.

### 11D Inadvertent Deviations

People may inadvertently disable alarms or interlocks, or they may deviate from procedure without understanding the full consequence of their actions. This may then result in accidents.

## 12 Constraints and Pressures

Each organisation has its own unique set of constraints that influence its management of safety and risk.

### 12A Lack of Resources

The organisation may not have the resources (money, manpower, etc.) to implement safety and invest in safety provisions, training, or equipment. If management feels that there are inadequate resources to manage the safety risks, they will often find other strategies to hide or transfer the risk.

### 12B Personal versus Organisational Resources

If the available resources are personal versus a company's, the decision maker may be reluctant to spend the resources. If the resources belong to the organisation rather than the decision maker, there will be more willingness to use it for safety, especially if the decision maker is accountable for safety performance.

### 12C Inadequate Time

When there is insufficient time to do something properly, shortcuts and improvisations are made, and accidents can result.

### 12D Delays and Liquidated Damages

Delays may occur to a project for various reasons, and liquidated damages may be payable for the delay. The contractor is more likely to take risks to rush the project.

### 12E Poor Planning and Scheduling

The lack of time to do something properly may be due to poor planning where insufficient time is allocated.

## 13 Circumvention and Dysfunction

Dysfunctions exist in organisations to different extent. Organizational dysfunction is the product of culture, structure, or leadership behaviours that undermines it management of safety.

### 13A Corruption of Function

The safety function may be corrupted by those parties that it is meant to control. If persons are given the authority to stop work or impose penalties on other parties, there is a potential for these other parties find ways to influence them not to do so. The corruption may also be in the form of allowing activities to proceed despite safety lapses or less stringent enforcement.

## 13B Voluntary and Involuntary Scapegoats

Scapegoats provide an opportunity for managers and organisations to take risk without having to face the consequences when things go wrong. Scapegoats may be voluntary, and they may receive benefits or payment for taking on the risk. This allows some stakeholders to take additional risks. Scapegoats are often required to put their names and signatures certifying that certain actions have been done or conditions fulfilled (e.g., permit to work, authorisations, approvals), but often there are inadequate resources to carry out those requirements.

They are often used when it is felt that an activity has to go on but there is no reasonable control measure that can prevent a serious accident from happening.

## 13C Blame Externalisation

Blame may be shifted or distributed to multiple parties so that causes and faults are harder to identify. If blame can be successfully distributed externally, certain risk behaviours will be more tolerable.

When much work is carried out by contractors, it is easy to shift blame to them when things go wrong.

This can also be achieved by obfuscating the causes of an accident by assigning blame to external factors

## 13D Fraud and Scams

Accidents may be created to claim certain benefits or insurance pay-outs. These accidents are typically preplanned or staged. Injuries or the severity of an injury in a real incident may be faked or injuries may even be self-inflicted. Uninsured losses are usually tolerable to the parties involved as they would be compensated for the insured losses.

### 13E Abuse of Safety Argument

Safety may be used as a reason to win arguments. This abuse of safety can affect the safety culture. For example, a manager can argue that a less desired option is less safe to justify his preferred option.

### 13F Office Politics

Like any other management function, safety may be affected by office politics. The usual office politics can degrade the effectiveness of the safety function.

One sad outcome of office politics is that competent and dedicated safety personnel may not be able to effectively perform their roles and functions.

Politically inept but professionally competent safety personnel may find themselves transferred out of departments or projects.

### 13G Suppressed Incident Reporting

Incident reporting is often suppressed for very good reasons (e.g., harsh punishments, disincentives). When incident reporting is suppressed, proper investigation may not be carried out, and learning from such incidents will not be possible. Management may be directly involved in the suppression, or they may react negatively to reports, thus discouraging any reporting.

Rewards and bonuses tied to accident-free statistics may result in the suppression of reporting where possible.

Workers may also refrain from reporting so as not to get their colleagues in trouble.

Different levels of management may suppress reporting in order to achieve their target of zero accidents.

## 13H Paid External Audits

Audits paid for by companies themselves will usually deliver the results required by customers, and thus these good results cannot be fully relied upon. Ways will be found to produce the desired result that is required by customers. For example, even if the management system is poorly implemented, the auditor may be easily satisfied with the evidence provided and the superficial compliance found.

Only a select few certification bodies in the world are able to maintain minimum standards for their audits.

It should also be noted that there are companies that are sincerely interested in improving the safety performance, but the auditors may not be competent enough to help them achieve their safety goals. Furthermore, certification auditors are prohibited from making any safety suggestions for improvements.

## 13I Forgery and Paperwork

It is not unusual to see forged or fabricated safety documents (e.g., PTWs, inspection checklists) from time to time. Paperwork such as procedures and records can also be created to fulfil safety requirements.

## 13J Defensive Safety Management

Defensive safety management is practised by management that is more concerned about being blameless in the event of an accident. They may make decisions that are irrational for the organisation but protect the self-interests of the managers involved, depleting resources. Often they do not care whether the safety measures are practical to comply as long as the onus is on someone else. This approach may extend to work by contractors and suppliers.

## 14 External Factors

Factors contributing to accidents may exist beyond the boundaries of organisations.

### 14A Uncontrollable Factors

There are uncontrollable factors which can contribute to accidents, such as weather, environmental factors, and natural disasters.

### 14B Unknown Causes

All accidents are the results of causes, but the causes of some accident may not be identifiable, and as such these accidents are hard to foresee and prevent.

Some accidents are extremely unlikely and unforeseeable, and this type of accidents is difficult to manage.

### 14C Business Environment

When the whole business environment is tough and uncertain, companies will be focused on battling these uncertainties, and safety may be the last thing on management's mind because survival is at stake. Digital disruption may reshape value chains and markets and transform the way business is conducted, rendering many practices obsolete and demanding new ways of safety management

An entire industry may undergo cycles, and during a down cycle, resources will be limited and safety management will be deprived of resources.

The business environment may also affect the supply and demand for good safety personnel, and at certain times it may be difficult to recruit good safety people.

Qualified contractors and subcontractors may be difficult to source in certain environment.

14D External Safety Function

The safety function may be outsourced to external organisations. These personnel performing the outsourced safety function have different levels of competence and motivation and they may not be dedicated to developing an effective and efficient safety management system in the company. Considerations by outsourced personnel can be short-term because they may not remain with the organisation.

External resources may not have the interests of the organisation in mind; often they will simply do the job that they are paid for in the most convenient way.

This issue is not totally external as an organisation can decide to perform this function internally.

# CHAPTER 5

# EFFICIENCY AND EFFECTIVENESS OF SAFETY MANAGEMENT SYSTEM

Efficiency without effectiveness misses the purpose,

Effectiveness without efficiency misses all the profit,

Effectiveness with efficiency brings performance.

—Martin Uzochukwu Ugwu

Ultimately whether a company continues to put resources into the safety management system depends on the benefits gained and management's perception of the value of the resulting loss prevention. If safety is seen as costing more money than the return it gives, then management motivation and support to drive safety will lessen. One should not expect management to have commitment when the system is wasting resources and delivering poor value. Thus, it is critical for companies to ensure that the safety management system is giving a good return on investment.

The efficiency and effectiveness of the safety management system is of prime importance if it is to continue to receive the continued support of management and business owners. Unfortunately, for various reasons many safety management systems continue to operate in ineffective or inefficient ways. With the right posturing and politics, management may even be praised for committing to an inefficient management system despite high costs expenditure. Some management may even embark on costly high-profile but largely ineffective safety programmes to demonstrate their safety commitment to stakeholders.

The real challenge in safety is not just to make a process safer, but to make that process safer with lesser resources. But reducing resources to a safety process when it is already optimal may cripple the effectiveness of that safety process so the right judgement is needed.

It is important to maintain the effectiveness of control measures while increasing efficiency (e.g., reducing the frequency of inspection).

The most efficient and effective way of managing safety is to inculcate a strong safety culture in the organisation. This requires much time and management consistency to build trust, as well as effort to set a strong platform for individuals to contribute and work towards safety.

## 5.1 Increasing Efficiency and Effectiveness

Efficiency is how much output (benefit) you get from the input (cost).

If input can be reduced to a minimum, efficiency will be increased.

Efficiency can be increased with the following.

- free safety
- Pareto principle
- technology
- reduction of paperwork
- removal of unnecessary controls
- reduction of frequency of control activity
- line ownership and involvement

- competition between suppliers and vendors
- volume discounts
- experience and familiarity

Effectiveness can be increased with the following.

- focus and customisation
- elimination of paper exercises
- avoiding the use of scapegoats
- "point of action" control

## 5.2 Free Safety

Safety is seldom totally free; otherwise, safety would no longer be a problem! There is often a cost to safety, but this cost may be minimised if well managed. Frequently, it may be possible to achieve safety with certain aspects of the costs reduced drastically to a very low or even zero cost.

A typical safety control measure would have the following aspects.

- cost of implementation
- material needed
- effort required
- time needed
- effectiveness

## Free Safety—Safety Culture

A strong safety culture results in employees wanting to work safely, even in the face of inadequate system, training, procedures, enforcement, and supervision. The most efficient way to achieve safety is to build and cultivate a strong safety culture in the organisation.

## Free Controls

Some safety control measures come free and are highly effective, such as avoiding "lines of fire" where energy may be released. Lines of fire can be identified, communicated to workers, and avoided. Communicating hazards to workers is also a cheap and easy way of preventing them from falling in harm's way.

## Free Time

Conduct safety training and briefings during operational lull periods.

Carry out certain safety activities at the end of shifts.

Safety awareness may be created during mealtimes or toilet breaks through message on posters, TV screens, and monitors.

## Free Material and Equipment

Materials and equipment used for safety may be recycled from another project or bought cheaply second-hand.

Materials may be repurposed and used for safety.

Unwanted equipment may be used for safety purposes.

Safety equipment may be constructed or manufactured internally, instead of being bought.

## Free Manpower

Leaders may be assigned secondary roles for championing various elements of the safety management system.

Employees may be asked to volunteer as emergency response team members or first-aiders.

Employees may be encouraged to form safety improvement teams to suggest safety ideas and improvements.

External organisations may be requested to provide free talks and training, especially those related to their safety services and products.

Interns may be hired to help with the safety management system.

Employees may be redirected to do safety tasks during production lull periods.

## Free Information and Knowledge

For every safety problem, usually someone has already found a solution. One simply needs to find and leverage it.

Much safety information can be obtained free from the Internet.

Free safety information may be obtained from authorities and industry groups.

Benchmarking visits may be made to similar organisations to get free learning.

## Free Resource

Occasionally, customers may be able to provide free resources or co-fund safety initiatives and incentives.

## Free Effort

If safety practices are nicely integrated with production activities, they may require little additional effort to execute. Similarly, paperwork or records may be integrated with production processes, thus incurring less cost, effort, and time.

Line supervisors supervising activities may easily integrate safety into their routine and thus incur little additional effort.

## Free Monitoring

Technology provides many solutions for monitoring the behaviour of workers through the use of cameras. Cameras can potentially reduce the number of safety enforcement personnel.

Cameras installed at strategic places can also assist in accident investigations.

The Internet may be used to transmit data and information remotely for monitoring purposes.

## Free Safety Management

Safety management software can be used to freely manage safety issues. Alerts are sent to responsible parties, information can be disseminated automatically, and data can be analysed.

Software can also be used to increase the efficiency of many safety processes (e.g., permit to work, management of change).

## Free Training

Many training programmes can be conducted at low cost over computer via e-learning. There is no need to pay for trainers or physical venues. Such training can also be taken outside of work hours.

### Free Ideas

Employees may be able to suggest free safety ideas. Employees may also be able to share practices and learning from their past companies.

Safety competitions may also be run to drive for free safety contributions, ideas, and creations from employees to work more safely.

Sharing visits to different companies may provide free exposure and ideas to improve safety.

### Free Contribution

Safety committees are a proven method to enable participation and contribution from employees because safety is mutually beneficial to workers' well-being.

### Free Culture

Employees may be appointed as safety advocates to help promote the safety culture.

## 5.3 Implementation Considerations

### Safety Culture

A strong safety culture is one of the most efficient and effective ways to implementing a safety management system. Even if the safety management system has deficiencies, safety culture may be able to overcome these weaknesses. In contrast, even if a safety management system is strong, a weak safety culture can find ways to undermine the system. The problem with safety culture is that it takes a long time to nurture, and constant resources are also required to grow it.

## Low-Hanging Fruits

The simplest and easiest control measures should be implemented first because these will produce the best and most efficient results. Although low-hanging fruits may not be the final, permanent solutions, they can produce the fastest risk-reduction results.

## Pareto Principle

The Pareto principle (also known as the 80/20 rule, the law of the vital few, or the principle of factor sparsity) states that for many events, roughly 80 per cent of the effects come from 20 per cent of the causes. If the most critical hazards can be controlled, the worst accident will be prevented in a company.

Seek first to control the most critical hazards to tolerable levels. Focus on the critical and refrain from wasting resources on the trivial.

Critical hazards are those that cause the greatest losses to the organisation.

## Technology

Technology can be used to increase the efficiency of safety management. This may include the following.

- Internet of things (IoT) and Internet of everything (IoE)
- cameras and CCTV
- artificial intelligence
- instantaneous communication
- data transmission and networking,
- Internet

- RFID, QR, and barcodes
- handheld devices
- biometric recognition
- GPS tracking devices
- sensors and detectors
- software programmes (e.g., for MOC, PTW)
- safety management software

**Reduction of Paperwork**

Other than serving as records, paperwork by itself typically does not reduce risk directly. However, adequate paperwork is still necessary to ensure the proper implementation of the safety management system.

**Removal of Unnecessary Controls**

Sometimes companies simply have to be seen as doing something, and so "something, anything" was implemented, but the activities do not actually reduce any risk. Unnecessary control measures should be minimised or eradicated because they use up time and resources with no real safety improvement.

Unnecessary safety management elements should not be implemented to maintain efficiency.

**Reduction of Frequency of Control Activity**

Frequency of control activities may be reduced or calibrated to actual needs such as

- inspection,
- audits, and
- refresher training.

Although the quantity of activities may be reduced, the quality of these activities must be maintained for effectiveness. When unnecessary control activities are carried out, people tend to go through the motion, wasting resources without any tangible results.

## Resisting Inefficient Safety Solutions

The easiest safety solutions that first come to mind usually make a work process slower and more cumbersome.

Overall efficiency will continue to drop as processes are made "safer and safer". The best safety solutions seek to maintain efficiency and productivity while reducing risk. There is thus a net gain instead of a loss by adopting these "efficient practices".

Management should not commit to continue implementing an inefficient and ineffective safety management system that contributes to losses instead of risk reduction.

An efficiency check should be carried out when developing safety control measures.

## Line Ownership and Involvement

If safety practices can be well integrated into the production process, the cost of implementation can be drastically reduced. When personnel get used to these requirements, these additional safety steps can be carried out habitually without incurring much cost or time.

## Competition between Safety Suppliers and Vendors

When a safety product becomes mainstream or an industry requirement, the cost of such products is often reduced because there will be many suppliers and vendors competing for the business.

## Volume Discounts

A volume discount is a financial incentive to encourage companies to purchase goods in large quantities. If a safety measure is adopted widely, the cost of it is likely to go down.

## Experience and Familiarity

With time and experience, a safety task can usually be performed more efficiently and with much reduced effort.

## Focus and Bespoke Safety Management System

Safety management systems need to be customised for different types of operations and activities. Here are some examples.

- process centred (e.g., process safety)
- people centred (e.g., personal safety)
- equipment centred (e.g., maintenance and integrity)
- material centred (e.g., DG management)
- hazard centred (e.g. control of hot work in confined space)
- industry specific (e.g., construction safety)

For maximum efficiency, a safety management system would normally have to be highly customised for an organisation, taking into consideration the unique situation of each organisation.

- types of activities
- nature of risk
- organisational culture
- size of organisation

## Elimination of Paper Exercises

Paper exercises give a false sense of security when ineffective controls are presented as functioning control measures.

## Avoiding the Use of Scapegoats

Although scapegoats serve the purpose of having the punitive consequences of accidents being transferred to them, such pseudoaccountability does not prevent accidents or lower risk. In fact, scapegoats shield others to take on more risks and absolve them of their responsibility and accountability.

Real, effective control measures should be developed instead of using scapegoats to escape the consequences.

## "Point of Action" Control

Control measures are only effective when they can change how work is actually being carried out. If there is no change to the way work is done at the point of action, all risk assessments, procedures, and control plans will be ineffective.

## Multiple Objectives per Activity

Safety activities cost time, money, and effort to implement. It is always best if they can kill a few birds with one stone. For example, a particular safety activity like general safety inspection can

- identify hazards,
- identify system deficiencies,
- suggest areas to improve quality or productivity,
- involve employees,
- show management's concern for safety,
- strengthen safety culture, and
- enhance safety management.

## 5.4 Unnecessary Risk

An unnecessary risk is risk that

- is avoidable,
- does not bring any profit or benefit, and
- can be easily controlled.

Unnecessary risks should be eliminated as a priority. Organisations and people can often be seen taking unnecessary risks (e.g., riding in a vehicle without wearing seatbelts).

Risks that can be easily eliminated should be eliminated.

# CHAPTER 6

## APPROPRIATE MANAGEMENT RESPONSES

> Courage consists, not in blindly overlooking danger, but in seeing and conquering it.
>
> —Jean Paul Richter

The following are suggested management responses to the various situations and factors described earlier. Circumstances will vary widely amongst organisations and their context, so the best solution will need to be customised for different organisations.

The factors and malaises are often interrelated, and a deeper causative factor may manifest itself in several ways. It may thus be more effective and efficient to address all issues rather than be limited to a focus on just the root causes.

A proper course of action or management approach may mitigate the effects of the factors and dysfunctions described earlier in Chapter 4. Some of these factors are easy to deal with (e.g. briefings, training) while others may be extremely difficult (e.g. profits, behavioural).

**Possible Strategies to Counter These Challenges**

1 Performance Targets and Profits Maximisation

1A No Profits for Safety

Safety should be integrated into the overall business strategy of the organisation and identified as a necessary requirement for the mission to achieve the organisation's vision. Safety should be aligned with the overall value chain so that it can be viewed as instrumental to the overall success of the business.

1B Profits and Profit Maximisation

Identify how safety can be aligned to business objectives and communicate this to key personnel.

Effort should be made to align safety objectives with overall business strategy (i.e., improving profits and business through improvements in safety performance and strong reputation in safety). Although immediate alignment with production objectives may not always be possible, companies should nonetheless try to find ways to use safety as competitive advantage, such as making clients aware of how improved safety performance will also benefit them in certain ways. Also, loss minimisation can contribute to greater profits.

Strong safety reputation can also help attract and retain talented employees.

1C Conflict of Objectives

Identify potential situations where conflict of objectives can arise.

Conflicts should be resolved holistically by aligning safety with the organisation's objectives. Resources and capabilities should be provided to management so that they are able to manage the safety risk to be best of their ability.

Management should set clear directions that safety should not be compromised and conflicts of objectives should be quickly resolved. In the short term, such decisions may affect profit and progress. An investigation should be made to understand how situations leading to such conflicts arise. Although such conflicts cannot be eliminated totally, it may be possible to find solution with tolerable risk. Conflicts can be pre-empted, and actions may be taken to mitigate and attenuate the conflicts.

Management should consistently provide unequivocal support for safety and find solutions that do not compromise safety.

1D Cost of Safety

Prepare a budget for safety.

Safety, when implemented judiciously, is a source of competitive advantage. Where possible, find ways to justify safety spending as an investment.

Try to find areas where there is "free safety". Free safety can drastically reduce the cost of safety initiatives.

Some clients may be willing to co-share some of the costs, especially if the need was unforeseen at the tender stage.

1F Cost vs. Benefits

Conduct cost benefit analysis for proposed safety control measures.

Safety measures should be effectively and efficiently implemented to obtain the greatest benefits with minimum costs. Continue to improve the effectiveness and efficiency of control measures.

The "as low as reasonably practicable" (ALARP) principal may be adopted to implement measures that are reasonably practicable.

## 2 Customer

Customers should show in interest in safety because it may benefit them, and a better managed company usually would be better able to deliver quality results as well.

### 2A No Desire for Safety

Engage and influence stakeholders over time.

Educate customers so that they are aware how improvement in safety can also benefit them as well, if possible. Find ways for customers to value safety in their service providers so that safety can become a competitive advantage.

### 2B Cheaper and Faster

Engage and advise customers that "cheaper and faster" can have implications on safety.

Customers, including developers, should not disregard safety in their rush for the completion of work or projects.

### 2C Not Paying for Safety

Identify ways to implement safety freely or more efficiently.

Only the most enlightened customers will be willing to copay for safety. Try to educate customers and explain to them the benefits and importance of safety. If customers show a preference for a safer service provider, it may be sufficient to tilt companies to focus more on safety.

Increase the efficiency of the system and identify more areas where "free safety" can be implemented.

### 2D Distractions

It is generally positive when customers are interested in safety, however this can degenerate into wastage of safety resources if customers are involved in an unconstructive manner. For example, a customer may ask for a trivial safety issue to be managed when critical safety issues have not been fully addressed. Customers should consider whether their involvement in safety is constructive. Customer should monitor the status of safety.

### 2E Fear of Involvement

When external parties can help to increase safety, they should get involved as much as reasonably practicable. Where possible, risks should be formally transferred to the organisation itself directly.

## 3 Competition

### 3A Lowest Quote

Identify ways to implement safety freely or more efficiently.

Ideally, the customer should also include safety as a criterion in awarding contracts and not simply award contracts to the lowest bid.

The customer should also prequalify bidders to ensure that they meet minimum safety performance, capability, and track record requirements. Customers should also ensure lowest bidders are aware of all the safety provisions in the contract specifications.

Safety management systems should be made more efficient so that safety costs may be lowered.

### 3B Unfair Competition

Look for ways to implement safety freely or more efficiently.

Ensure that all service providers and contractors abide by minimum safety requirements. Inform stakeholders if there is an uneven playing field. Unsafe companies can save costs by taking on risks and cutting back on safety measures. Although this may reduce costs in the short term, potential risks may be very high, and workers may be exposed to unacceptable risks.

### 3C Ultracompetition

To avoid the problem of companies cutting safety personnel or safety provisions, contractual requirements should be clearly and comprehensively stated. It should also be made known to suppliers and contractors that checks will be made to ensure compliance.

### 3D Bidding Mistakes

If the lowest bid is much lower than the second lowest bid, a clarification meeting should be held to ascertain whether all provisions will be catered to.

## 4 Requirements

### 4A Lack of External Requirements, Legal or Other

Engage and influence stakeholders over time.

All parties in positions of power, including customers, should establish requirements for safety, check on compliance, and monitor safety implementation. Interest and involvement should be shown. All parties can ultimately benefit from improvements in safety performance.

Communicate to stakeholders the benefits of improved safety performance and the need for practical safety requirements.

### 4B Inadequate Checks and Enforcement

Engage and influence stakeholders over time.

Periodic compliance checks and audits should be made to ensure compliance.

### 4C Certification and Superficial Compliance

Although it may be good to require suppliers and contractors to be certified, companies should not rely solely on certification as an assurance that service providers are adequately managing their safety risks. Checks, visits, and audits should be made to important suppliers.

### 4D Inadequate Internal Requirements and Standards

Organisations should establish internal requirements and standards for their operations. If internal requirements are well established and communicated, managers will plan for compliance as early as possible to lower the cost of implementation.

### 4E Unrealistic Requirements

Help stakeholders to understand the actual situation.

Safety requirements should be reasonable, realistic, practical and attainable. The cost of safety implementation needs to be paid from the profits from the activities. If the cost of implementing a safety requirement exceeds profits, this will not be sustainable.

### 4F Superficial Requirements

Stakeholders should scrutinise requirements to check that they are sufficient to reduce the risk as needed. Requirements must drive change to the way work is being carried out such that it is safer.

### 4G Ambiguous Requirements

Requirements should be clearly stated for ease of compliance. Interpretations should be given publicly if they are not.

### 4H Hollow Buzzwords

Management should minimise the use of hollow buzzwords and empty words. Talk should be supported by action.

## 5 Implementation Efficiency and Effectiveness

### 5A Paper exercises and Going through the Motion

Ineffective practices should be reduced, and effectiveness should be increased.

Requirements should be meaningful, purposeful, and realistic so that companies are motivated to comply with the requirements. Workers should be made aware of the importance of the effort required. Checks should be made on the effectiveness of controls implemented. Effectiveness evaluations should be periodically carried out.

## 5B Disconnected and Poorly Designed Management Systems

A plan to review the effectiveness of each critical element of the safety management system should be established.

Personnel developing the safety management system should be experienced and competent. System evaluation may be carried out, and cross-audits should be conducted.

The safety management system must be connected to the point of action and be based on the risk exposure of the company.

## 5C Ineffective/Inefficient Control Measures and Poor Fixes

Ineffective practices should be reduced, and effectiveness should be increased.

Changes should make the process not only safer but also better. Good control measures are those that increase safety while maintaining quality and productivity (e.g., simply making an activity slower may increase safety, but it will negatively impact productivity).

## 5D Progressive Reduction in Efficiency

Maintain and increase efficiency of the safety management system.

The effort to increase the efficiency and effectiveness of the safety management system should be an ongoing process. Redundancy may be reduced to increase efficiency.

Refrain from hijacking the safety management system by adding other concerns to it. The safety management system must not lose its focus on safety. Integrate other concerns (e.g., quality, health, environment, security) only when the safety portion has been effectively implemented.

5E Depleting Resources on Noncritical Hazards

The Pareto principle should be observed where relevant and resources should be directed at critical safety issues. Safety resources should not be wasted on noncritical issues. Prioritisation of safety issues should be carried out rationally and based on risk level.

5F "Flavour of the Month" Safety

The safety management system should be established based on real needs and risk. Implementation should be carried out consistently instead of being abandoned for more fashionable programmes.

5G Negative Process

Management should turn safety can be turned into a positive process through the following:

- giving recognition and praise for good effort
- competition and prizes
- celebrating good performance
- motivational communication
- rewarding good safety performance

- showing care and concern for each other
- organising health-related activities

### 5H Lack of Compliance

A system of checks or audits should be implemented for all critical control measures to ensure compliance.

Unless an activity is profitable by itself, it may not be carried out if there are no checks and consequences. All relevant personnel must know that checks and audits on critical items will be periodically carried out.

### 5I Ultrasafety

Risk management should be logical, and benefits should be weighed against costs. Resources should be rationally and judiciously allocated. Risk assessment should generally be accurate. Resources should go to strengthen the weakest areas and not be frittered away unnecessary in strong or pet areas.

### 5J False Indications

Conclusions about the state of safety management should be made based on correct and appropriate indicators.

## 6 Human

### 6A Errors, Slips, Lapses, Mistakes

Processes may be improved such that chance or human errors are minimised. "Error-proofing" efforts should be initiated for all critical activities.

Checklists should be used to prevent lapses. A "pointing and calling" system may be implemented to increase concentration and reduce mistakes.

### 6B Laziness, Easier Ways, Rushing, and Shortcuts

Situations where a deviation from procedure can cause catastrophic consequences should be identified and avoided.

Management must monitor and supervise employees to ensure compliance to procedures and safety requirements. Workers should be reminded to follow procedures and not take shortcuts or rush their work during pre-work briefings.

### 6C Worker Risk Attitudes and Complacency

Workers may have different risk attitudes, and some will be more inclined to take risks. Training on risk attitudes may be given to workers so that they are more aware of their own risk attitudes and better able to refrain to making irrational risk decisions and taking unnecessary risk.

Workers should be periodically warned against complacency. Relevant incidents in other places should be shared to workers so that they are aware of the possible consequences. Individually, workers may normalise risk over time.

Inspections should also look out for and correct complacent behaviours.

### 6E Resistance to Change

Management should recognise the need for safety improvement even if past practices have not resulted in accidents before.

## 6D Safety Creep

Various audits (internal, external, corporate, cross, etc.) may be carried out to check for consistency in practices and compliance to internal requirements. "Cold eye" reviews of existing practices may be conducted.

A strong "management of change" or "internal audit" system may also be able to catch some issues. Cross-audits (with a sister site) may also be useful. Creep needs to be eradicated at the ground level.

## 6F Habit and Practices

It is important for management to pre-empt unsafe behaviours when there is a change or quickly identify unsafe behaviours before they become a habit or practice. Effort must be put into eradicating any unsafe behaviour before they take root.

## 6G Overload, Confusion, Distractions, and Fatigue

Systems should be designed such that people are not overloaded with information. Information should be clearly presented so that people are not confused.

Sources of distractions (e.g., handphones) should be disallowed or removed where possible.

Checks should be made that workers have the minimum level of mindfulness and concentration for the task at hand.

## 6H Foreign Cultural Differences

Programmes and initiatives developed with foreign cultural elements should be evaluated, and modified if necessary, to the local culture to make them more effective.

6I Improper Motivation and Production Incentives

Management should refrain from tying incentives solely to production targets. Faster but unsafe production behaviour should not be encouraged or tolerated. Management should monitor if workers are improperly motivated to take risks. Management should not penalise workers for working more slowly for the sake of safety.

6J Fatalism

Generally, fatalists will still comply with the safety management system requirements because they can believe that they are fated to be in their current position and thus need to do what is already predetermined. It is only when they are given the choice to do things in different ways that they may choose a riskier path.

Organisations should allow religious practices at the workplace because people usually believe that the gods love them and will protect them. Workers with fatalistic beliefs should be monitored closely to confirm that they are working safely and in accordance with procedure.

Workers need to believe that their safe behaviours are needed to prevent accidents from happening.

7 Management

7A Inadequate Control and Management

Inadequate control and management is usually due to management's inexperience in managing safety or a lack of resources. Management should gain awareness of the most cost-effective safety measures to have a competitive advantage. Management should expose themselves more to industry practices and adopt those that are beneficial. Management must ensure that operational control is maintained. Adequacy of control must be monitored and evaluated continuously by top management and staff safety function.

The safety management system should rest on and be supported by the quality management system. All inputs to processes should meet requirements in order to have a minimal level of assurance. If people are not trained or equipment improperly maintained, then safety management by itself will be futile.

### 7B Lack of Safety Management Experience

Both formal and informal training in safety management should be given to all levels of management. Management with safety responsibilities should be given more specific safety training and orientation.

### 7C Inadequate Safety Function

Organisations should seek to find talent in safety management. Safety professionals should be qualified in terms of experience, training, education, and qualifications. Safety departments should be adequately staffed.

### 7D Avoiding Self-Blame and Aversion to Criticism and Deficiencies

Organisations should establish a non-fault-finding culture when things go wrong. Genuine mistakes should be forgiven so that people can be less defensive and more open to improvements and preventing recurrence of accidents. A "no blame" policy should be implemented for certain issues.

Management should respond positively to criticisms of the safety management system and constructive suggestions. They should be open to ideas for improvement and learn from better companies.

### 7E Supervisory Conflict

Managers should evaluate whether supervisors are able to enforce safety requirements in a fair and consistent manner.

Line supervisors should be held accountable for the safety performance of workers. Supervisors should be required to report any serious safety violations.

### 7F Friends and Colleagues

Higher management should periodically evaluate line compliance with safety requirements and whether requirements are effectively enforced. Listen to any complaints about consistency and fairness of enforcement.

When safety personnel have a strong relationship with other employees, it can be a strength when safety initiatives need to be implemented.

### 7G Risk Perception

Management should be given some training in risk and safety management.

### 7H Counterintuitive Effects

Managers should be informed of possible counterintuitive effects in safety management.

### 7I Safety—A Staff Function Responsibility

Safety should not be managed solely by the safety department. All workers should be involved, and line departments should be accountable for safety performance and have ownership of safety.

Internal safety function should be advisory in nature.

### 7J Focus on Lagging Indicators and Equate "No Accident" as Safety

Organisations should not overly focus on lagging indicators and treat the absence of accident as safety. The safety management system should continue to generate safe behaviours regardless of accident data.

Organisations should avoid rewarding "no accidents". Tie rewards to safety activities, effort, or contributions.

It should be understood that "no accident" simply means that an accident has not happened during the period and may not mean that safety is well managed.

## 8 Organisation

### 8A High-Risk Appetite and Risk-Taking Culture

Management should communicate policies clearly on safety demonstrate leadership.

### 8B Poor Safety Culture

Management at all levels should

- communicate value of safety,
- provide safety leadership,
- set safety expectations,
- allocate safety responsibilities and accountabilities,
- increase awareness of hazards and safe behaviours,
- respond to safety issues and incidents appropriately,
- allocate resources to safety and take actions,
- rewards and incentives for safety contributions,
- monitor effectiveness of safety management system,
- build trust through fairness,
- build teamwork in safety, and
- celebrate safety successes.

Leaders need to consistently cultivate the safety culture.

A strong safety culture would include the following.

- leadership and teamwork
- trust and respect
- communication and reporting
- ownership
- care and concern

- personal responsibility

- awareness and knowledge

- learning and improvement

## 8C Power Dysfunction and Safety Leadership

It is rare for senior managers to hold themselves accountable for accidents unless they are held accountable by regulators. However, it may also be true that some operational decisions leading to accidents are made by lower level managers and supervisors.

All levels of management should provide leadership for safety. Top management should be visibly involved in safety.

## 8D Lack of Line Ownership

Hold line management accountable for safety performance, incidents, and safety improvements. Safety function should perform more of an advisory role than mainly enforcement.

## 8E Risk Habituation and Normalisation of Risk

Managers and workers should be warned of the dangers of risk normalisation. Managers from various locations should cross visit each other's sites to check that workers are not taking risk for granted.

## 8F Business Risk and Risk Aversion

People need to recognise the difference between taking business risk versus operational risk.

Risk aversion to risk may also translate into implementing unnecessary draconian control measures. Such actions reduce the overall effectiveness of the safety management system and may also reduce support for the system. Risk control decisions should be rational and commensurate with the risk.

8G Dogma and Political Incorrectness

Discussions on safety should be based on facts, data, and rationale.

8H Cyclical Commitment

Management's effort in safety should not be reliant on external events or recent incidents. Effort should be continuous and consistent to drive the safety culture and maintain maximum effectiveness of the system.

8I Poor Safety Organisational Structure

Proper organisation structure should be in place. Safety function should report to a higher level of management that can make balanced decisions with accountability for safety performance. The safety function requires a degree of independence to be able to perform its role effectively.

8J Lack of Talent

Management should attract good candidates to work in the safety function. The company can consider rotating key safety personnel out of the safety function, if necessary, for career progression. Management should also turn safety into a positive process from a negative process so that capable people may be attracted to the profession.

## 9 Industry

When an industry adopts a good set of safety requirements, usually companies are able to implement these requirements more cost-effectively because some companies are able to find the most efficient ways to implement them, and the supply of products and services are more competitive because there is a bigger market and volume.

### 9A Lack of Industry Standard

Good cost-effective practices should be shared in the industry. Standards should be set to continually raise industry standards incrementally. Industry groups should establish realistically achievable industry standards.

### 9B Inadequate Sharing of Good Practices

Platforms to share good practices should be established, and benchmarking should be facilitated.

### 9C Inadequate Sharing of Incident Learning

Channels for sharing of incidents and near misses should be established. Incident sharing can generate a state of vigilance and eradicate complacency.

### 9D Inappropriate Industry Standards for Safety

Industry standards for safety should be sustainable for the industry. Relevant stakeholders should be consulted so that the safety standard set is realistic and achievable. Capability building and assistance programs should be implemented if necessary, to help industry comply with the standard. The preeminent stakeholder(s) should lead the consultation process and plan for the long-term success of the new standard.

## 9E Criteria for Safety Awards

Safety awards should not be based on accident statistics. Safety award criteria should be based on things like

- safety effort,
- safety knowledge,
- safety contribution, and
- safety ideas and innovation.

## 9F Safety as a Market Entry Barrier

Check to ensure safety requirements are effective in improving safety.

## 9G External Funding and Subsidies

A safety management system must be able to make a business case for itself. A safety management system must be able to provide results, returns, and benefits right from the start. A simple system, fine-tuned over time, can produce quick results and effective results for companies. As a company gains experience with implementing the system, its efficiency and effectiveness will increase. Companies can then consider how to grow the system to increase their coverage and overall effectiveness.

It is important for companies to avoid creating paper exercises that produce no long-term results.

Properly allocated resources can result in real improvements in companies' safety management (e.g., capability building, development of tools).

### 9H Subcontractors and Sub-Subcontractors

The developer or owner representatives should ensure that the main contractor exercise full control over subcontractors and sub-subcontractors and be fully accountable.

Clients should monitor the appointment of subcontractors as different groups of workers bring in a different safety culture.

Subcontractors' performance should be monitored at site.

Subcontractors should be clear about all contractual requirements for safety.

## 10 Complexity and Nature of Accidents

### 10A Technology

Appropriate tools (e.g., Hazop, What-if) should be used to analyse and better understand the hazards from technology.

### 10B Infinite Possibilities

> The problem with the future is that more things might happen than will happen.
>
> —Plato

Appropriate tools (e.g., Hazop, What-if) should be used to analyse and identify myriad possibilities.

Not all possibilities need to be eliminated, only those with intolerable consequences that can result in a fatality. These critical few possibilities need to be eliminated. A robust hazard analysis process should be used to identify as many possibilities as possible. Possibilities should be eliminated with proper design and engineering controls. Hierarchy of control should be used wherever possible.

Management of change procedure may be needed to control changes with new possibilities.

If possible, allocate a little effort to make worst-case scenarios slightly more tolerable or survivable.

### 10C Probability, Chance, and Unpredictability

Absence of incident does not mean that a process is safe. Constant effort should be put into continual improvement of safety. Areas for continual improvement should be identified.

### 10D Safety Outcomes Not Proportional to Safety Effort

Establish a set of good leading indicators so that safety effort can be measured. Effort and resources put into safety should be consistent, and not only when there are accidents. As a priority, high-potential or intolerable risk exposures should be eliminated.

### 10E Murphy's Law

Whatever that can go wrong should be identified, and if the risk is not tolerable, it should be eliminated.

### 10F Flawed Investigation and Root Cause Analysis

> There are a thousand hacking at the branches of evil to one who is striking at the root.
>
> —Henry David Thoreau

Investigators for serious accidents must be properly trained and skilful in obtaining evidence to arrive at the real, underlying, contributory and root causes of an accident.

Root cause analysis often stops at deficiencies in the overall management system without going further into what caused those deficiencies. Accident investigations should go wider and deeper to identify all possible causes through causal analysis. Preventive actions should be taken to eliminate as many causes as possible.

## 10G Knee Jerks after Accidents

Management should be aware of and refrain from counterproductive knee-jerk reactions. Causes of accidents and management deficiencies should be thoroughly investigated, and proper corrective and preventive actions should be instituted.

Stakeholders should refrain from blame allocation and focus more on accident prevention.

## 11 Natural State of Ignorance

### 11A Workers

Safety induction training is the first step followed by other generic and specific safety training. Ongoing toolbox meetings and communication should be conducted by leaders.

### 11B Management

Management should have some basic training in safety management. Management should also attend industry sharing events and discuss safety issues with safety staff function.

### 11C Customers

Customers should be informed on improvements in safety and environmental impact at the production stage that may also benefit them or others. Advise customers on the ways they can be meaningfully involved in safety. Inform customers of the consequences on safety, if any, should they make unreasonable demands. Establish industry practices so that customers have to abide by certain acceptable practices that do not adversely impact safety.

### 11D Inadvertent Deviations

People should be made aware of the potential consequences of deviations from procedures. A stronger management of change (MOC) procedure may be required. Inspections should also identify whether there have been any unauthorised changes to procedures and conditions.

Controls should be implemented to ensure that bypassing of interlocks and disabling of alarms are done safely with controls and not casually without considerations.

## 12 Constraints and Pressures

### 12A Lack of Resources

Identify ways to implement safety freely or more efficiently. Management should find ways to secure necessary resources prior to commencement of activities.

### 12B Personal versus Organisational Resources

Resources for safety should be budgeted and utilised as necessary. Resources should be provided if there are proper justifications.

### 12C Inadequate Time

Adequate time should be allocated to ensure that activities can be carried out safely. Customers should be advised against unrealistic project schedules and delivery times.

### 12D Delays and Liquidated Damages

If delays are not due to the fault of the contractor, they should not be required to pay liquidated damages. Project completion schedule should be revised.

### 12E Poor Planning and Scheduling

Proper planning, scheduling, and preparation should be enforced so that activities can be carried out smoothly and safely.

## 13 Circumvention and Dysfunction

### 13A Corruption of Function

Depending on circumstance, some anticorruption measures may need to be implemented (e.g., rotation of personnel). There should be internal channels for reporting and whistle-blowing.

### 13B Voluntary and Involuntary Scapegoats

If there is a desire for risks to be managed, there should be recognition of potential scapegoat situations and eliminating such situations. The real decision makers and owners of operations should be identified.

### 13C Blame Externalisation

Critical responsibilities should be clearly assigned so that blame cannot be easily shifted away.

### 13D Fraud and Scams

Ensure existing arrangements do not motivate workers or companies to create fake accidents.

Establish a process to investigate suspicious accidents. Train personnel to investigate such incidents.

Because such acts are criminal, there are already penalties. Record-keeping should be thorough so that all activities are traceable. CCTVs may also be installed at strategic locations so that relevant evidence may be captured. Investigations into fraud and scams should be robust so that the truth may be uncovered.

### 13E Abuse of Safety Argument

Safety may be used as an argument in company discussion when this real objection is not safety related. Top management should recognise this and step into the discussion if necessary.

### 13F Office Politics

Politics may affect the proper functioning and effectiveness of the safety function. Top management should recognise this and manage such politics.

### 13G Suppressed Incident Reporting

Organisations should systematically identify and remove barriers to incident reporting. Reporting of incidents should be encouraged. Management should respond positively to incident reports and avoid blame allocation as far as possible.

Management should not in any way punish the bearer of bad news.

### 13H Paid External Audits

The party that is interested in the audit results should arrange the audit or appoint the auditors. If the auditor is paid by the auditee, then the audit results would somehow have to meet the requirements of the auditee.

### 13I Forgery and Paperwork

Management should not request or require workers to forge safety documents and records.

### 13J Defensive Safety Management

Top management should avoid developing a blame culture so that middle managers and safety personnel do not defensively manage safety.

## 14 External Factors

### 14A Uncontrollable Factors

Although uncontrollable factors may not be controlled, they may be detectable or predictable. Proper planning and design may be able to overcome some of these issues (e.g., wet season could be avoided, lightning conditions).

The risk from these conditions may also be controlled or mitigated (e.g., lightning protection system, flood protection).

## 14B Unknown Causes

Accidents have different degrees of complexity, preventability, and foreseeability. Some accidents are hard to prevent because their causes may be unknown or unforeseeable. Unusual or less conventional methods may need to be used to identify and resolve such causes.

Strong emergency responses can reduce the potential losses of a major event.

Business continuity management (BCM) can make processes more resilient, and further losses may be mitigated because BCM does not care what causes the disruption.

## 14C Business Environment

Organisations should try to be consistent in its safety management, and standards should be upheld as far as possible even in difficult times.

## 14D External Safety Function

Ideally, the safety function should be performed by employees motivated to establish an effective and efficient safety management system within the company.

# CHAPTER 7

# LOSS CONTROL MATRIX

> The best strategy is always to be very strong, first generally, then at the decisive point.
>
> —Carl von Clausewitz

The Loss Control Matrix (LCM) consists of forty-nine elements organised into a framework of seven levels (1–7) and seven groups (A–G). Each group of elements focuses on a different aspect of safety management:

- analysis
- behaviours
- culture
- defences
- equipment
- facility
- general

Levels 1–3: Basic elements

Levels 4–5: Intermediate elements

Levels 6–7: Advanced elements

**Progressive Approach**

The safety management system must always be able to deliver results and values that support the business. As a guide, the lower level elements should be implemented first.

Level 1 and 2 elements are the more basic elements that typically produce quick results and risk reduction benefits when efficiently implemented.

The implementation of the loss control elements should be improved for efficiency and effectiveness before going to higher levels and more difficult elements.

Elements should be implemented progressively from level 1 onwards until level 5. Elements of a comprehensive safety and health management system can all be found within level 5. Beyond level 5 are the elements that are applicable to organisations with special needs and circumstances, where a higher level of control and performance is needed.

A certified system is considered complete and meeting all the requirements from day one that the system is certified. Further major change is rarely seen in these certified systems after certification unless there is a major nonconformity or change in standard.

**Selection of Relevant Elements for Implementation**

Elements relevant to each organisation should be identified and evaluated for implementation where applicable. Not all the elements need to be implemented unless they are relevant. If an element is applicable, it should be considered for implementation eventually.

For an Occupational Health and Safety (OH&S) Management System to deliver value, it needs to be appropriate to the risk exposure of the organisation. Because organisations have different risk exposures, the system implemented need to be customised for maximum effectiveness and efficiency. The programmes and systems should be implemented only if they are relevant and suitable (i.e., effective and efficient) to the organisation, and they should not all be blindly implemented. Unlike the ISO requirements in ISO 45001, not all the programmes and systems listed should be implemented because they may not be suitable.

Irrelevant elements should be dropped. Elements that do not deliver the requisite results should be more frequently evaluated for improvements.

Some elements may passively reduce risk whereas other elements do so actively.

Organisations that have been certified to ISO 45001 should also be able to declare the number of elements here that have been effectively implemented.

As a minimum, all level 1 elements should be implemented in every organisation.

In lieu of the elements listed in the LCM, organisations may choose to replace them with equivalent or more relevant programmes that can deliver greater results.

## Safety Strategy

The first target should be to increase the effectiveness of the system implemented.

The next target would then be to improve the efficiency of the system. When these two targets have been achieved, the overall safety would also have improved in a more sustainable manner.

It is better to have a few strong elements than many weak elements implemented.

A senior leader should be appointed to champion individual elements of the safety management system.

## Advantages of the Element Approach

The element approach holds several advantages.

- Safety management work can be compartmentalised into manageable size.

- Persons can be assigned to each relevant element.

- Division of labour through appointment of element champions for different elements

- Element can be more easily managed because the element's scope can be defined, and effectiveness can be more easily evaluated.

- The system is scalable and expanded as required.

- Focus and resources can be directed at key elements.

- Implementation can be progressive with future new elements.

- The complexity of system implemented is commensurate with the risk.

## Disadvantages of the Element Approach

Disadvantages include the following.

- Links between elements may be obscured or not understood.

- Multiple elements to control a single hazard may not be properly coordinated.

- There may be delays in implementing a new required element.

## Common Requirements in Each Element Program

Each element of the system should be managed with its own "Plan, Do, Check, Act" continual improvement cycle. The following requirements should be included in the implementation of each element or into the "Element Process Steps" described in each element, where relevant.

1. roles, responsibilities, and authorities (ISO 45001 Clause 5.3)
2. system and element requirements (who/what/when/how often) (Clause 4.4)
3. planning (Clause 6)
4. OH&S objectives (Clause 6.2.1)
5. specific legal and other requirements (Clause 6.1.3)
6. documented information (Clause 7.5)
7. operational planning and control (Clause 8.1)
8. monitoring, measurement, analysis, and performance evaluation (Clause 9.1)
9. evaluation of compliance (Clause 9.1.2)
10. internal audit (Clause 9.2)
11. management review (Clause 9.3)
12. improvement (Clause 10)

Analysis

    A1  Risk Assessment

    A2  Hazard Analysis

    A3  Engineering and Design Analysis

    A4  Training Needs Analysis

    A5  Root Cause Analysis

    A6  Behaviour Analysis

    A7  Data Analysis and Review

Behaviour and Personal

    B1  Safety Induction Programme and Toolbox Meetings

    B2  General Safety Rules and Life-Saving Rules

    B3  Compliance and Enforcement System

    B4  Competence Assurance System

    B5  Communication and Coordination Meetings

    B6  Behaviour Safety Observation

    B7  Personal Safety Contact

Culture

    C1  Employee Participation

    C2  Safety and Health Committee and Subcommittees

    C3  Safety and Health Promotion

- C4 Safety and Health Awards and Recognition
- C5 Hazards, Concern, Near Miss, and Whistle-Blower Reporting System
- C6 Leadership
- C7 Safety Culture

## Defences

- D1 PPE Programme
- D2 High-Risk Operations Control Programme
- D3 Hazardous Chemicals Control Programme
- D4 Occupational Health Control Programmes
- D5 Hygiene Monitoring and Medical Surveillance Programme
- D6 Ergonomics and Fatigue Management
- D7 Individual Risk Factors and Management

## Equipment

- E1 Tools, Equipment, and Critical Parts Inspections
- E2 Maintenance System
- E3 Machine Guarding and Automation Safety
- E4 Lockout Tag-Out Programme and De-Energisation
- E5 Asset Integrity Programme
- E6 Control System Safety

E7 Safety in Procurement

Facility

F1 General Safety and Health Inspection Programme

F2 Signs, Colour-Coding, Labelling, and Tagging System

F3 Housekeeping, Order, and Cleanliness Programme

F4 Management of Change and Pre-Startup Safety Review

F5 Capital Projects Safety Management

F6 Regulations, Codes, and Internal Standards

F7 Safety Information System

General

G1 Procedures and Safe Work Practices

G2 Emergency Preparedness and Crisis Management

G3 Permit to Work System

G4 Contractor Management Programme

G5 Investigation and Learning from Incident System

G6 Leading Indicators and Monitoring

G7 New Technology, Improvement, and Benchmarking

Table—Loss Control Matrix

| LCM | 1 | 2 | 3 | 4 | 5 | 6 | 7 |
|---|---|---|---|---|---|---|---|
| **Analysis** | Risk Assessment | Hazard Analysis | Engineering and Design Analysis | Training Needs Analysis | Root Cause Analysis | Behaviour Analysis | Data Analysis and Review |
| **Behaviour** | Induction and Toolbox | Safety Rules and Life-Saving Rules | Compliance and Enforcement | Competence Assurance | Communication and Coordination | Behaviour Task Observation | Personal Safety Contact |
| **Culture** | Employee Participation | Safety Committee | Safety Promotion | Awards and Recognition | Reporting System | Leadership | Safety Culture |
| **Defences** | Personal Protective Equipment | High-Risk Operation | Hazardous Chemical | Health Control | Hygiene Monitoring Surveillance | Ergonomics and Fatigue | Individual Risk Factors |
| **Equipment** | Tools and Equipment Inspection | Maintenance System | Machine Guarding Automation | Energy Controls and LOTO | Reliability and Integrity | Control System Safety | Procurement |
| **Facility** | General Inspection | Signs, Colour Coding | Housekeeping, Order, and Cleanliness | MOC and PSSR | CAPEX | Regulations Codes and Standards | Safety Information System |
| **General** | Procedure | Emergency Preparedness | PTW System | Contractor Management | Investigation and LFI System | Leading Indicators | Technology Improvement Benchmark |

The most applicable and relevant elements should be implemented first. The effectiveness of an element is more important than mere implementation. It is more important to implement a few elements effectively than many elements ineffectively.

The lower level elements are generally the low-hanging fruits that should receive priority in implementation. They should be implemented first because

1. they are usually easier to implement,
2. they are easier to increase in effectiveness,
3. they usually require less resources,
4. they are likely to have a higher cost-benefit ratio, and
5. they are the fastest to deliver results.

Generally, higher level elements should not be implemented before the lower level elements are implemented because

1. basic issues have not been resolved,
2. management or workforce is not ready,
3. the system is still too immature, and
4. higher level elements are costly to implement and may not produce any results if not effectively implemented.

# CHAPTER 8

# ELEMENTS OF THE LOSS CONTROL MATRIX

> To sustain an environment suitable for man, we must fight on a thousand battlegrounds.
>
> —Lyndon Baines Johnson

A brief description of each element in the Loss Control Matrix (LCM) is given below.

Where available, relevant requirements of ISO 45001 and references in the ISO 45001 Annex A are extracted. As such, the numbering of the clauses will not be sequential because only the more relevant points are extracted.

Relevant requirements and references are in bold for ease of reference.

Information contained in ISO 45001 Annex A are not requirements, and certification auditors do not issue nonconformance notes against them.

## A—ANALYSIS

> No problem can be solved from the same level of consciousness that created it.
>
> —Albert Einstein

## A1. Risk Assessment

*"Identify hazards"*

### Element Description:

Risk assessment is the process to identify hazard and risk factors, evaluate the probability and severity, and determine the control measures.

Risk assessment includes:

- identification of hazards and risk factors that have the potential to cause harm (hazard identification)

- analysis and evaluation of the risk associated with that hazard (risk analysis, and risk evaluation)

- elimination of the hazard or control of the risk when the hazard cannot be eliminated (risk control)

Risk assessment should cover all routine and non-routine activities.

Risk assessments can:

- increase workers' awareness of hazards

- identify persons at risk

- determine risk control measures

- reduce risk at an early stage

- facilitate compliance with legal requirements

**Element Process Steps:**

1. establish risk assessment procedure
2. establish an inventory of all activities
3. carry out risk assessment (identify hazard, evaluate risk)
4. review and approve risk assessment
5. communicate results of risk assessment
6. implement risk control measures
7. evaluate effectiveness and improve

**ISO 45001 Requirements:**

**6.1.2 Hazard identification and assessment of risks and opportunities**

**6.1.2.1 Hazard identification**

The organisation shall establish, implement and maintain a process(es) for hazard identification that is ongoing and proactive.

**6.1.2.2 Assessment of OH&S risks and other risks to the OH&S management system**

The organisation shall establish, implement and maintain a process(es) to:

a) assess OH&S risks from the identified hazards, while taking into account the effectiveness of existing controls;

b) determine and assess the other risks related to the establishment, implementation, operation and maintenance of the OH&S management system.

The organisation's methodology(ies) and criteria for the assessment of OH&S risks shall be defined with respect to their scope, nature and timing to ensure they are proactive rather than reactive and are used in a systematic way. Documented information shall be maintained and retained on the methodology(ies) and criteria.

### ISO 45001 Annex A:

### A.6.1.2 Hazard identification and assessment of risks and opportunities

### A.6.1.2.1 Hazard identification

The ongoing proactive identification of hazard begins at the conceptual design stage of any new workplace, facility, product or organisation. It should continue as the design is detailed and then comes into operation, as well as being ongoing during its full life cycle to reflect current, changing and future activities.

While this document does not address product safety (i.e. safety to end-users of products), hazards to workers occurring during manufacture, construction, assembly or testing of products should be considered.

Hazard identification helps the organisation recognize and understand the hazards in the workplace and to workers, in order to assess, prioritize and eliminate hazards or reduce OH&S risks.

Hazards can be physical, chemical, biological, psychosocial, mechanical, electrical or based on movement and energy.

## A2. Hazard Analysis

*"Analyse hazards"*

**Element Description:**

Depending on the industry and processes, the appropriate hazard analysis methodologies should be identified (e.g., JHA/JSA and/or PHA) to provide a deeper understanding the risks involved and the possible mitigation.

PHA methodologies may include hazard and operability study (HAZOP), What-If, FMEA, and more. PHA should be carried out for covered and other high-risk processes.

Job hazard analysis (JHA) or job safety analysis (JSA) should be carried out systematically with the identification of jobs and tasks to be analysed. JHA should be carried out on critical tasks as a priority. Critical tasks are tasks which have the potential to cause a serious accident if they are not performed properly.

Job hazard analysis consists of the following:

- break the job task into steps
- identify the hazard of each step
- identify ways to eliminate or reduce the hazard

Process hazard analysis (PHA) should be revalidated periodically or at least every five years. PHA should be reviewed every three to five years or when there are significant changes or incidents.

**Element Process Steps:**

(Example for JHA)

1. identify critical tasks for hazard analysis
2. analyse each critical task
3. break critical task down into steps
4. eliminate or control hazard
5. communicate JHA
6. evaluate effectiveness and improve

**ISO 45001 Requirements:**

**6.1 Actions to address risks and opportunities**

**6.1.1 General**

The organisation, in its planning process(es), shall determine and assess the risks and opportunities that are relevant to the intended outcomes of the OH&S management system associated with changes in the organisation, its processes or the OH&S management system.

**6.1.2.1 Hazard identification**

The organisation shall establish, implement and maintain a process(es) for **hazard identification** that is ongoing and proactive. The process(es) shall take into account, but not be limited to:

b) routine and non-routine activities and situations, including hazards arising from:

1) infrastructure, equipment, materials, substances and the physical conditions of the workplace;

**6.1.2.2 Assessment of OH&S risks and other risks to the OH&S management system**

The organisation shall establish, implement and maintain a process(es) to:

a) assess OH&S risks from the **identified hazards**, while taking into account the effectiveness of existing controls;

b) determine and assess the other risks related to the establishment, implementation, operation and maintenance of the OH&S management system.

The organisation's **methodology(ies)** and criteria for the assessment of OH&S risks shall be defined with respect to their scope, nature and timing to ensure they are proactive rather than reactive and are used in a systematic way. Documented information shall be maintained and retained on the methodology(ies) and criteria.

**ISO 45001 Annex A:**

**A.6.1 Actions to address risks and opportunities**

**A.6.1.1 General**

OH&S opportunities address the identification of hazards, how they are communicated, and the **analysis** and mitigation of known hazards. Other opportunities address system improvement strategies.

Examples of opportunities to improve OH&S performance:

b) job **hazard analysis** (job safety analysis) and task-related assessments;

### A.6.1.2.2 Assessment of OH&S risks and other risks to the OH&S management system

An organisation can use different methods to assess OH&S risks as part of its overall strategy for addressing different hazards or activities. The method and complexity of assessment does not depend on the size of the organisation, but on the hazards associated with the activities of the organisation.

Other risks to the OH&S management system should also be assessed using appropriate methods.

### A3. Engineering and Design Analysis

*"Engineering"*

**Element Description:**

Design and engineering reviews or analyses cover a wide scope of structures, equipment, and machinery that may be carried out to proactively eliminate hazards at the design and engineering stage and ensure that they are fit for purpose.

Different types of systems may have different methodologies and approaches to the analysis.

Techniques used for such reviews may include the following.

- checklists
- engineering calculations,
- FOS (Factor of Safety)
- DFS (Design for Safety) Reviews
- FMEA (Failure Modes and Effects Analysis)

- Fault Tree Analysis
- Reliability Engineering and System Safety
- SIL (Safety Integrity Level) Assessment
- Facility Siting
- QRA (Quantitative Risk Assessment)
- Ergonomic/Human Factor Analysis

Analysis techniques may be qualitative or quantitative. One objective is to find causal dependencies between a hazard on a system level and failures of individual components.

## Element Process Steps:

1. identify situations where a design review or analysis is required
2. establish procedure or checklists for review/analysis
3. identify personnel to carry out reviews/analysis
4. train personnel to carry out reviews/analysis
5. establish procedure to manage design and engineering processes
6. evaluate effectiveness and improve

## ISO 45001 Requirements:

### 6.1.2.1 Hazard identification

The organisation shall establish, implement and maintain a process(es) for hazard identification that is ongoing and proactive. The process(es) shall take into account, but not be limited to:

f) other issues, including consideration of:

1) the **design of work areas, processes, installations, machinery/equipment**, operating procedures and work organisation, including their adaptation to the needs and capabilities of the workers involved;

### 6.1.4 Planning action

The organisation shall take into account the hierarchy of controls (see 8.1.2) and outputs from the OH&S management system when planning to take action.

### 8.1.2 Eliminating hazards and reducing OH&S risks

The organisation shall establish, implement and maintain a process(es) for the elimination of hazards and reduction of OH&S risks using the following hierarchy of controls:

a) eliminate the hazard;

b) substitute with less hazardous processes, operations, materials or equipment;

c) use engineering controls and reorganisation of work;

### ISO 45001 Annex A:

### A.6.1.2.1 Hazard identification

The ongoing proactive identification of hazard begins at the conceptual design stage of any new workplace, facility, product or organisation. It should continue as the design is detailed and then comes into operation, as well as being ongoing during its full life cycle to reflect current, changing and future activities.

## A4. Training Needs Analysis

*"Train people"*

**Element Description:**

The training system may include the following.

- training needs analysis
- training matrix
- training plan
- evaluation of training effectiveness

Training needs should be identified to ensure competence of personnel.

Relevant legal requirements for OH&S training must be identified.

OH&S training should be identified. OH&S training may include:

- chemical safety
- use of PPE
- firefighting

**Element Process Steps:**

1. identify skills and competencies required

2. inventory of training needs
3. identify/develop training programmes
4. develop training plan
5. implement training plan
6. evaluate training effectiveness and improve

**ISO 45001 Requirements:**

**5.4 Consultation and participation of workers**

The organisation shall establish, implement and maintain a process(es) for consultation and participation of workers at all applicable levels and functions, and, where they exist, workers' representatives, in the development, planning, implementation, performance evaluation and actions for improvement of the OH&S management system.

The organisation shall:

e) emphasize the participation of non-managerial workers in the following:

> 4) determining competence requirements, **training needs,** training and evaluating training (see 7.2);

**7.2 Competence**

The organisation shall:

a) determine the necessary competence of workers that affects or can affect its OH&S performance; b) ensure that workers are competent (including the ability to identify hazards) on the basis of appropriate education, **training** or experience;

b) where applicable, take actions to acquire and maintain the necessary competence, and evaluate the effectiveness of the actions taken;

c) retain appropriate documented information as evidence of competence.

Note Applicable actions can include, for example, the provision of **training** to, the mentoring of, or the reassignment of currently employed persons, or the hiring or contracting of competent persons.

### 8.1.2 Eliminating hazards and reducing OH&S risks

The organisation shall establish, implement and maintain a process(es) for the elimination of hazards and reduction of OH&S risks using the following hierarchy of controls:

d) use administrative controls, including **training;**

### ISO 45001 Annex A:

### A.8.1.2 Eliminating hazards and reducing OH&S risks

The hierarchy of controls is intended to provide a systematic approach to enhance occupational health and safety, eliminate hazards, and reduce or control OH&S risks. Each control is considered less effective than the one before it. It is usual to combine several controls in order to succeed in reducing the OH&S risks to a level that is as low as reasonably practicable.

The following examples are given to illustrate measures that can be implemented at each level.

d) Administrative controls including **training**: conducting periodic safety equipment inspections; conducting training to prevent bullying and harassment; managing health and safety coordination with subcontractors' activities; conducting induction training; administrating forklift driving licences; providing instructions on how to report incidents, nonconformities and victimization without fear of retribution; changing the work patterns (e.g. shifts) of workers; managing a health or medical surveillance programme for workers who have been identified as at risk (e.g. related to hearing, hand-arm vibration, respiratory disorders, skin disorders or exposure); giving appropriate instructions to workers (e.g. entry control processes).

## A5. Root Cause Analysis

*"Identify root causes"*

### Element Description:

The purpose of root cause analysis is to identify the factors (immediate and basic causes) that have resulted in an incident to determine systemic causes, behaviours, actions, inactions, or conditions that need to be changed to prevent recurrence of similar incidents and to identify lessons that may be learnt to prevent other related incidents.

Root causes are defined as factors that if corrected, would prevent recurrence of the incident. The root causes typically identified are system deficiencies, management failures, etc.

Corrective actions undertaken without thorough root cause analysis (RCA) may not address root causes and may only deal with the symptoms of deeper problems. Corrective actions should target every stage of the accident causation chain, where possible.

Causal factors are defined as the events and conditions that produce or contribute to an incident. Causal factors include

- direct causes,
- contributing causes, and
- root causes

A suitable root cause analysis methodology should be selected for use in the organisation. Overly complicated methodologies may not be suitable because line personnel will have difficulties using them.

Root cause analysis should be carried out with

- proven RCA methodologies,
- training of relevant personnel, and
- procedural requirements for RCA.

### Element Process Steps:

1. establish procedure for RCA (select RCA methodology)
2. train relevant personnel to use methodology
3. conduct RCA
4. evaluate effectiveness and improve

### ISO 45001 Requirements:

### 10.2 Incident, nonconformity and corrective action

The organisation shall establish, implement and maintain a process(es), including reporting, investigating and taking action, to determine and manage incidents and nonconformities.

b) evaluate, with the participation of workers (see 5.4) and the involvement of other relevant interested parties, the need for corrective action to eliminate the **root cause(s)** of the incident or nonconformity, in order that it does not recur or occur elsewhere, by:

   1) investigating the incident or reviewing the nonconformity;

   2) determining the cause(s) of the incident or nonconformity;

**ISO 45001 Annex A:**

**A.6.1 Actions to address risks and opportunities**

**A.6.1.1 General**

OH&S opportunities address the identification of hazards, how they are communicated, and the analysis and mitigation of known hazards. Other opportunities address system improvement strategies.

Examples of opportunities to improve OH&S performance:

e) incident or nonconformity investigations and corrective actions;

Examples of other opportunities to improve OH&S performance:

— enhancing the incident investigation process(es);

**A.10.2 Incident, nonconformity and corrective action**

**Root cause analysis** refers to the practise of exploring all the possible factors associated with an incident or nonconformity by asking what happened, how it happened and why it happened, to provide the input for what can be done to prevent it from happening again.

When determining the **root cause** of an incident or nonconformity, the organisation should use methods appropriate to the nature of the incident or nonconformity being analysed. The focus of **root cause analysis** is prevention. This analysis can identify multiple contributory failures, including factors related to communication, competence, fatigue, equipment or procedures.

## A6. Behaviour Analysis

*"Understand behaviours"*

### Element Description:

Based on the behaviour safety observation programme and feedback from supervisors and workers, unsafe behaviours in an organisation can be identified and analysed.

Behaviour analysis is conducted on unsafe behaviour (i.e., unsafe acts, at-risk behaviours) to understand the reasons why workers behave in these unsafe ways. Factors that drive specific unsafe behaviours and the consequences to the worker's behaviour are identified in the analysis.

Resistance and barriers to safe behaviours should be reduced or removed. Causes (e.g., workplace factors) and motivation (e.g., consequences and rewards) for unsafe behaviour should be eliminated or reduced. Workers' needs and expectations should be determined in the behaviour analysis.

Positive behaviour reinforcement techniques should be used to reinforce safe behaviours.

Antecedents should be introduced to initiate the desired behaviour.

All levels of management should be trained in giving safe behaviour reinforcement.

**Element Process Steps:**

1. identify problem or target behaviours
2. conduct behaviour analysis
3. identify interventions
4. evaluate effectiveness and improve

**ISO 45001 Requirements:**

**4.2 Understanding the needs and expectations of workers and other interested parties**

The organisation shall determine:

a) the other interested parties, in addition to workers, that are relevant to the OH&S management system;

b) the relevant **needs and expectations** (i.e. requirements) **of workers** and other interested parties;

**ISO 45001 Annex A:**

**A.10.2 Incident, nonconformity and corrective action**

Separate processes may exist for incident investigations and nonconformities reviews, or these may be combined as a single process, depending on the organisation's requirements.

**Root cause analysis** refers to the practice of exploring all the possible factors associated with an incident or nonconformity by asking what happened, how it happened and why it happened, to provide the input for what can be done to prevent it from happening again.

When determining the root cause of an incident or nonconformity, the organisation should use **methods appropriate** to the nature of the incident or nonconformity being analysed. The focus of root cause analysis is prevention. This **analysis** can identify multiple contributory failures, including factors related to communication, competence, fatigue, equipment or procedures.

### A7. Data Analysis and Review

*"Analyse data"*

**Element Description:**

Data from various safety programmes can be analysed:

- results of risk assessment
- inspection findings
- accident investigation findings
    - types of accidents
    - activity or operation
    - equipment involved
    - time
    - injured part
    - age

- length of service
- immediate causes
- root causes
- near miss investigation findings
- results of root cause analysis
- hazard reporting data
- employee feedback data
- BBS observation data

Results of safety data analysis can be used to:

- identify common causes of accidents
- identify hotspots
- develop control strategy and/or interventions
- develop safety goals and objectives
- develop leading indicators
- identify safety programmes and projects
- refine and improve safety programmes

Pareto analysis may be used to analyse results, outcomes, and causes.

Relevant software and hardware will be needed to collect and process the data.

Results of data analysis should be periodically presented to senior management.

**Element Process Steps:**

1. identify data to be analysed
2. collect data
3. analyse data
4. identify actions
5. evaluate effectiveness and improve

**ISO 45001 Requirements:**

**9.1 Monitoring, measurement, analysis and performance evaluation**

**9.1.1 General**

The organisation shall establish, implement and maintain a process(es) for monitoring, measurement, analysis and performance evaluation.

The organisation shall determine:

a) what needs to be monitored and measured, including:

1) the extent to which legal requirements and other requirements are fulfilled;

2) its activities and operations related to identified hazards, risks and opportunities;

3) progress towards achievement of the organisation's OH&S objectives;

4) effectiveness of operational and other controls;

b) the methods for monitoring, measurement, analysis and performance evaluation, as applicable, to ensure valid results;

c) the criteria against which the organisation will evaluate its OH&S performance;

d) when the monitoring and measuring shall be performed;

e) when the results from monitoring and measurement shall be analysed, evaluated and communicated.

The organisation shall evaluate the OH&S performance and determine the effectiveness of the OH&S management system.

## 9.3 Management review

Top management shall review the organisation's OH&S management system, at planned intervals, to ensure its continuing suitability, adequacy and effectiveness.

The management review shall include consideration of:

d) **information on the OH&S performance, including trends in**:

1) incidents, nonconformities, corrective actions and continual improvement;

2) **monitoring and measurement results;**

3) results of evaluation of compliance with legal requirements and other requirements;

4) audit results;

5) consultation and participation of workers;

6) risks and opportunities;

e) adequacy of resources for maintaining an effective OH&S management system;

g) opportunities for continual improvement.

The outputs of the management review shall include decisions related to:

— continual improvement opportunities;

— any need for changes to the OH&S management system;

— resources needed;

— actions, if needed;

— opportunities to improve integration of the OH&S management system with other business processes;

— any implications for the strategic direction of the organisation.

**ISO 45001 Annex A:**

**A.9.1 Monitoring, measurement, analysis and performance evaluation**

**A.9.1.1 General**

**Analysis is the process of examining data** to reveal relationships, patterns and trends. This can mean the use of statistical operations, including information from other similar organisations, to help draw conclusions from the data. This process is most often associated with measurement activities.

**Performance evaluation** is an activity undertaken to determine the suitability, adequacy and effectiveness of the subject matter to achieve the established objectives of the OH&S management system.

## B. Behaviour

### B1. Safety Induction and Toolbox Meetings

*"Talk safety to people"*

**Element Description:**

The purpose of the safety induction program is to inform every relevant worker of the safety requirements and OH&S hazards so that they can work safely.

The safety induction program may include:

- safety and health policy
- safety rules and regulations
- PTW system
- PPE requirements
- emergency response
- incident reporting

A toolbox meeting is a safety meeting for workers that focuses on safety topics related to the specific job and typically includes:

- jobs or tasks to be performed

- OH&S hazards, risk assessments conducted
- work procedures and practices
- safety requirements and checks to be carried out

**Element Process Steps:**

1. identify information to be communicated
2. establish communication plan and schedule
3. develop communication materials
4. communicate information
5. check understanding
6. evaluate effectiveness and improve

**ISO 45001 Requirements:**

**5.1 Leadership and commitment**

Top management shall demonstrate leadership and commitment with respect to the OH&S management system by:

    e) communicating the importance of effective OH&S management and of conforming to the OH&S management system requirements;

**7.3 Awareness**

Workers shall be made aware of:

c) the implications and potential consequences of not conforming to the OH&S management system requirements;

e) hazards, OH&S risks and actions determined that are relevant to them;

f) the ability to remove themselves from work situations that they consider present an imminent and serious danger to their life or health, as well as the arrangements for protecting them from undue consequences for doing so.

## 7.4 Communication

### 7.4.1 General

The organisation shall establish, implement and maintain the process(es) needed for the internal and external communications relevant to the OH&S management system, including determining: a) .........

### **ISO 45001 Annex A:**

**A.8.1 Operational planning and control**

**A.8.1.1 General**

Operational planning and control of the processes need to be established and implemented as necessary to enhance occupational health and safety, by eliminating hazards or, if not practicable, by reducing the OH&S risks to levels as low as reasonably practicable for operational areas and activities.

Examples of operational control of the processes include:

g) adapting work to workers; for example, by:

2) the induction of new workers;

### A.8.1.2 Eliminating hazards and reducing OH&S risks

The following examples are given to illustrate measures that can be implemented at each level.

    d) Administrative controls including training: conducting periodic safety equipment inspections; conducting training to prevent bullying and harassment; managing health and safety coordination with subcontractors' activities; **conducting induction training;** administrating forklift driving licences; providing instructions on how to report incidents, nonconformities and victimization without fear of retribution; changing the work patterns (e.g. shifts) of workers; managing a health or medical surveillance programme for workers who have been identified as at risk (e.g. related to hearing, hand-arm vibration, respiratory disorders, skin disorders or exposure); giving appropriate instructions to workers (e.g. entry control processes).

## B2. General Safety Rules and Life-Saving Rules

*"Rules"*

### Element Description:

General safety rules regulate the general behaviour of workers at the workplace and usually define the minimum standards that must be followed by every worker.

Typical general safety rules cover:

- general PPE requirements
- compliance to procedures

- housekeeping
- photography and video-taking
- reporting of incidents
- emergency evacuation
- alcohol and drugs
- inappropriate behaviours (e.g., horseplay)
- safe transportation

Life-saving rules are rules that, when not followed, may result in serious injuries or fatality. Organisations may also call them critical rules, cardinal rules, or noncompromising rules. Normally, an organisation should have zero tolerance for any violation of life-saving rules because they can lead to a fatal accident. Life-saving rules should be kept to a minimum and effectively communicated to all workers and contractors.

Typical life-saving rules apply to:

- LOTO
- work in confined spaces
- PTW compliance
- work at height
- lifting operations
- driving safety
- hot work

**Element Process Steps:**

1. identify needs for safety rules
2. consult employees
3. establish rules
4. implement rules
5. enforce rules
6. evaluate rule effectiveness and improve

**ISO 45001 Requirements:**

**8.1.2 Eliminating hazards and reducing OH&S risks**

The organisation shall establish, implement and maintain a process(es) for the elimination of hazards and reduction of OH&S risks using the following hierarchy of controls:

d) use administrative controls, including training;

**ISO 45001 Annex A:**

**A.8.1 Operational planning and control**

**A.8.1.1 General**

Operational planning and control of the processes need to be established and implemented as necessary to enhance occupational health and safety, by eliminating hazards or, if not practicable, by reducing the OH&S risks to levels as low as reasonably practicable for operational areas and activities.

Examples of operational control of the processes include:

a) the use of procedures and **systems of work;**

**A.9.1 Monitoring, measurement, analysis and performance evaluation**

**A.9.1.1 General**

In order to achieve the intended outcomes of the OH&S management system, the processes should be monitored, measured and analysed.

c) Examples of what could be monitored and measured to evaluate the fulfilment of other requirements can include, but are not limited to:

   1) collective agreements (when not legally binding);

   2) standards and codes;

   3) corporate and other policies, **rules and regulations;**

**B3. Compliance and Enforcement System**

*"Check compliance"*

<u>**Element Description:**</u>

In any organisation, there is a possibility of non-compliance to safety requirements by groups or individuals. The goal of an effective disciplinary programme is primarily to discourage workers from behaving in a manner that would put themselves or others at risk.

A system for compliance checks should be implemented to monitor compliance and enforce safety requirements. A fair and consistent enforcement system should be in place to manage minor and major non-compliances to safety requirements. This process should include opportunities for behavioural improvement, such as:

- coaching
- counselling
- retraining

An investigation should be carried out if any disciplinary action is taken against a worker.

Typical steps of disciplinary actions:

- verbal warning
- written warning
- suspension
- termination

There should be a formal appeal process and this process should be documented and communicated to workers.

Workers should be informed of:

- relevant legal requirements for worker's safe work and conduct
- disciplinary process for major violation

If relevant, a system for managing contractors' violations should also be implemented.

A system to recognise and reward compliance may also be implemented if it is appropriate to the working environment.

Small tokens or recognition may be awarded to workers for their compliance.

## Element Process Steps:

1. establish procedure for compliance and enforcement
2. establish disciplinary system
3. implement system of enforcement checks
4. evaluate effectiveness and improve

## ISO 45001 Requirements:

### 7.3 Awareness

Workers shall be made aware of:

c) the implications and potential consequences of not conforming to the OH&S management system requirements;

### 9.1.2 Evaluation of compliance

The organisation shall establish, implement and maintain a process(es) for evaluating compliance with legal requirements and other requirements (see 6.1.3).

### 10.2 Incident, nonconformity and corrective action

The organisation shall establish, implement and maintain a process(es), including reporting, investigating and taking action, to determine and manage incidents and nonconformities.

### ISO 45001 Annex A:

### A.9.1.2 Evaluation of compliance

The frequency and timing of compliance evaluations can vary depending on the importance of the requirement, variations in operating conditions, changes in legal requirements and other requirements and the organisation's past performance. An organisation can use a variety of methods to maintain its knowledge and understanding of its compliance status.

### A.10.2 Incident, nonconformity and corrective action

Separate processes may exist for incident investigations and nonconformities reviews, or these may be combined as a single process, depending on the organisation's requirements.

### B4. Competence Assurance System

*"Competent workers"*

### Element Description:

Workers must be competent to perform tasks that, if performed wrongly, may result in serious accidents.

Competence criteria should be established for critical positions. Criteria should be based on:

- education
- training
- experience

Components of competence would include:

- knowledge
- skill
- ability

Competencies are observable, measurable, and verifiable.

Assessment and verification of competence levels should be carried out.

Workers performing critical activities may need to be licensed.

A safety passport system documenting training attended may be useful in certain industries.

Competence assurance may include:

- competence framework
- competency standards
- assessment process
- personal development plans

Benefits of a competence assurance programme are:

- enhanced safety assurance
- defined standards for acceptable performance
- skilled workforce
- gaps in knowledge, skills, and abilities are identified
- higher productivity and quality

### Element Process Steps:

1. identify all positions requiring competence
2. establish competence criteria
3. competency assessment
4. plan to close competency gaps
5. evaluate effectiveness and improve

### ISO 45001 Requirements:

### 5.4 Consultation and participation of workers

The organisation shall:

e) emphasize the participation of non-managerial workers in the following:

4) determining competence requirements, training needs, training and evaluating

training (see 7.2);

### 7.2 Competence

The organisation shall:

a) determine the necessary competence of workers that affects or can affect its OH&S performance;

b) ensure that workers are competent (including the ability to identify hazards) on the basis of appropriate education, training or experience;

## 8.1.2 Eliminating hazards and reducing OH&S risks

The organisation shall establish, implement and maintain a process(es) for the elimination of hazards and reduction of OH&S risks using the following hierarchy of controls:

d) use administrative controls, including training;

### ISO 45001 Annex A:

### A.8.1 Operational planning and control

### A.8.1.1 General

Operational planning and control of the processes need to be established and implemented as necessary to enhance occupational health and safety, by eliminating hazards or, if not practicable, by reducing the OH&S risks to levels as low as reasonably practicable for operational areas and activities.

Examples of operational control of the processes include:

b) ensuring the competence of workers;

## B5. Communication and Coordination Meetings

*"Know"*

### Element Description:

Incompatible activities, when carried out simultaneously, can result in serious accidents. Proper coordination is required to avoid such situations.

Coordination is especially important when several groups of people are working at the same location or in proximity.

Safety information needs to be communicated to all relevant parties in a timely manner.

Information to be communicated includes:

- hazards
- results of risk assessment and hazard analysis
- safe work practices and procedures
- rules and regulations
- PPE requirements
- safety plans

**Element Process Steps:**

1. identify coordination needs
2. establish coordination requirements
3. implement coordination meetings and requirements
4. evaluate effectiveness and improve

**ISO 45001 Requirements:**

**7.4 Communication**

**7.4.1 General**

The organisation shall establish, implement and maintain the process(es) needed for the internal and external **communications** relevant to the OH&S management system, including determining:

a) on what it will **communicate**;

### 7.4.2 Internal communication

The organisation shall:

a) internally **communicate** information relevant to the OH&S management system among the various levels and functions of the organisation, including changes to the OH&S management system, as appropriate;

b) ensure its **communication** process(es) enables workers to contribute to continual improvement.

### 8.1 Operational planning and control

### 8.1.1 General

At multi-employer workplaces, the organisation shall **coordinate** the relevant parts of the OH&S management system with the other organisations.

### 8.1.4.2 Contractors

The organisation shall **coordinate** its procurement process(es) with its contractors, in order to identify hazards and to assess and control the OH&S risks arising from:

a) the contractors' activities and operations that impact the organisation;

b) the organisation's activities and operations that impact the contractors' workers;

### 8.1.4.3 Outsourcing

The organisation shall ensure that outsourced functions and processes are controlled. The organisation shall ensure that its outsourcing arrangements are consistent with legal requirements and other requirements and with achieving the intended outcomes of the OH&S management system. The type and degree of control to be applied to these functions and processes shall be defined within the OH&S management system.

NOTE **Coordination** with external providers can assist an organisation to address any impact that outsourcing has on its OH&S performance.

### ISO 45001 Annex A:

### A.8.1.2 Eliminating hazards and reducing OH&S risks

The following examples are given to illustrate measures that can be implemented at each level.

d) Administrative controls including training: conducting periodic safety equipment inspections; conducting training to prevent bullying and harassment; **managing health and safety coordination with subcontractors' activities;** conducting induction training; administrating forklift driving licences; providing instructions on how to report incidents, nonconformities and victimization without fear of retribution; changing the work patterns (e.g. shifts) of workers; managing a health or medical surveillance programme for workers who have been identified as at risk (e.g. related to hearing, hand-arm vibration, respiratory disorders, skin disorders or exposure); giving appropriate instructions to workers (e.g. entry control processes).

### A.8.1.4.2 Contractors

The need for **coordination** recognizes that some contractors (i.e. external providers) possess specialized knowledge, skills, methods and means.

An organisation can achieve **coordination** of its contractors' activities through the use of contracts that clearly define the responsibilities of the parties involved. An organisation can use a variety of tools for ensuring contractors' OH&S performance in the workplace (e.g. contract award mechanisms or prequalification criteria which consider past health and safety performance, safety training, or health and safety capabilities, as well as direct contract requirements).

When **coordinating** with contractors, the organisation should give consideration to the reporting of hazards between itself and its contractors, controlling worker access to hazardous areas, and procedures to follow in emergencies. The organisation should specify how the contractor will coordinate its activities with the organisation's own OH&S management system processes (e.g. those used for controlling entry, for confined space entry, exposure assessment and process safety management) and for the reporting of incidents.

### B6. Behaviour Task Observation Programme

*"See and talk"*

**Element Description:**

The purpose of a behaviour task observation program is to identify any unsafe behaviour and hazards and to change people's unsafe behaviours into safe behaviours.

Critical tasks are tasks which have the potential to cause a serious accident when not performed properly. Critical tasks should be included in the observation programmes.

Behaviour-based safety programmes may be implemented to strengthen safety culture and compliance.

Typical activities in BBS programmes include:

- training of observers
- observation targets
- establish observation checklist
- analysis of observation data

**Element Process Steps:**

1. establish behaviour observation programme
2. identify behaviour standards
3. identify critical tasks to observe
4. train observers
5. Implement BBS programme
6. collect and analyse data
7. evaluate effectiveness and improve

An observation programme empowers peers to comment and intervene on instances of unsafe acts by their fellow workers.

## ISO 45001 Requirements:

**3.30**

**monitoring**

determining the status of a system, a process (3.25) or an activity

Note 1 to entry: To determine the status, there may be a need to check, supervise or critically observe.

**5.4 Consultation and participation of workers**

The organisation shall:

d) emphasize the consultation of non-managerial workers on the following:

7) determining what needs to be monitored, measured and evaluated (see 9.1);

## ISO 45001 Annex A:

**A.9.1 Monitoring, measurement, analysis and performance evaluation**

**A.9.1.1 General**

Monitoring can involve continual checking, supervising, critically observing or determining the status in order to identify change from the performance level required or expected. Monitoring can be applied to the OH&S management system, to processes or to controls. Examples include the use of interviews, reviews of documented information and observations of work being performed.

## B7. Personal Safety Contact

*"One to one"*

### Element Description:

A personal safety contact programme is a one-on-one contact in which an interaction between worker and a mentor is made to discuss safety issues. The safety contact is tailored to the context of the individual worker and relevant issues specific to the worker. Personal, sensitive, and confidential safety-related issues may be discussed in such an environment.

Supervisors should encourage workers to speak up on any safety-related issues or concerns.

Personal safety contacts should be planned in advance following a schedule.

The objectives of a planned safety contact programme are:

- communicate a specific safety message or requirement

- reinforce personal safety performance, commending good practices

- resolve personal safety performance and motivational issues

- address any recent unsafe behaviours

- obtain feedback from the worker on any work-related safety issues

PSC programmes may include:

- training on PSC

- PSC targets
- planned PSC schedule
- PSC records

**Element Process Steps:**

1. identify personnel requiring PSC
2. establish personal safety contact programme
3. set PSC standards
4. train supervisors
5. implement PSC programme
6. collect and analyse data
7. evaluate effectiveness and improve

**ISO 45001 Requirements:**

**3.5**

**consultation**

seeking views before making a decision

**5.1 Leadership and commitment**

Top management shall demonstrate leadership and commitment with respect to the OH&S management system by:

l) ensuring the organisation establishes and implements a process(es) for consultation and participation of workers (see 5.4);

## 5.4 Consultation and participation of workers

The organisation shall establish, implement and maintain a process(es) for consultation and participation of workers at all applicable levels and functions, and, where they exist, workers' representatives, in the development, planning, implementation, performance evaluation and actions for improvement of the OH&S management system.

The organisation shall:

a) provide mechanisms, time, training and resources necessary for consultation and participation;

Note 1 Worker representation can be a mechanism for consultation and participation.

c) determine and remove obstacles or barriers to participation and minimize those that cannot be removed;

Note 2 Obstacles and barriers can include failure to respond to worker inputs or suggestions, language or literacy barriers, reprisals or threats of reprisals and policies or practises that discourage or penalize worker participation.

d) emphasize the consultation of non-managerial workers on the following:

   1) determining the needs and expectations of interested parties (see 4.2);

## ISO 45001 Annex A:

### A.5.4 Consultation and participation of workers

The consultation and participation of workers, and, where they exist, workers' representatives, can

be key factors of success for an OH&S management system and should be encouraged through the

processes established by the organisation.

Consultation implies a two-way communication involving dialogue and exchanges. Consultation involves the timely provision of the information necessary for workers, and, where they exist, workers' representatives, to give informed feedback to be considered by the organisation before making a decision.

### A.6.1 Actions to address risks and opportunities

### A.6.1.1 General

Examples of other opportunities to improve OH&S performance:

— improving the process(es) for worker consultation and participation;

### A.9.1 Monitoring, measurement, analysis and performance evaluation

### A.9.1.1 General

In order to achieve the intended outcomes of the OH&S management system, the processes should be monitored, measured and analysed.

a) Examples of what could be monitored and measured can include, but are not limited to:

1. **occupational health complaints**, health of workers (through surveillance) and work environment;

2. work-related incidents, injuries and ill health, and **complaints,** including trends;

Monitoring can involve **continual checking, supervising**, critically observing or determining the status in order to identify change from the performance level required or expected. Monitoring can be applied to the OH&S management system, to processes or to controls. Examples include the use of **interviews**, reviews of documented information and **observations of work being performed.**

## C. Culture

## C1. Employee Participation

*"Involvement"*

### Element Description:

An employee involvement programme provides a mechanism for workers to be directly involved in protecting their own safety and health. Workers frequently are the most knowledgeable about the details of the operations, and often they are also the ones most exposed to the hazards.

Employee involvement is a key component of cultivating a strong safety culture. Worker involvement programmes provide for:

- worker empowerment
- consultative process

- open communication
- mutual trust
- management response to workers' feedback

Workers may be involved in:

- risk assessment
- identifying competence requirements and training needs
- incident investigation and prevention
- promotional activities
- total productive maintenance
- safety improvement projects
- contribution of safety ideas and suggestions

Workers should be appointed, or they may volunteer as "safety advocates". Safety advocates

- influence co-workers to work safely
- identify and highlights safety issues at work
- provide safety information

### Element Process Steps:

1. identify areas where employees can participate
2. engage employees for their participation
3. monitor employee participation results

4. evaluate effectiveness and improve

**ISO 45001 Requirements:**

3.4

participation

involvement in decision-making

Note 1 to entry: **Participation** includes engaging health and safety committees and workers' representatives, where they exist.

**5.1 Leadership and commitment**

Top management shall demonstrate leadership and commitment with respect to the OH&S management system by:

l) ensuring the organisation establishes and implements a process(es) for consultation and **participation** of workers (see 5.4);

**5.2 OH&S policy**

Top management shall establish, implement and maintain an OH&S policy that:

> f) includes a commitment to consultation and **participation** of workers, and, where they exist, workers' representatives.

**5.4 Consultation and participation of workers**

The organisation shall establish, implement and maintain a process(es) for consultation and **participation** of workers at all applicable levels and functions, and, where they exist, workers' representatives, in the development, planning, implementation, performance evaluation and actions for improvement of the OH&S management system.

The organisation shall:

>  a) provide mechanisms, time, training and resources necessary for consultation and **participation;**

NOTE 1 Worker representation can be a mechanism for consultation and **participation.**

>  b) provide timely access to clear, understandable and relevant information about the OH&S management system;
>
>  c) determine and remove obstacles or barriers to **participation** and minimize those that cannot be removed;

NOTE 2 Obstacles and barriers can include failure to respond to worker inputs or suggestions, language or literacy barriers, reprisals or threats of reprisals and policies or practises that discourage or penalize worker **participation.**

>  d) emphasize the consultation of non-managerial workers on the following:
>
>  >  1) determining the needs and expectations of interested parties (see 4.2);
>  >
>  >  2) establishing the OH&S policy (see 5.2);
>  >
>  >  3) assigning organisational roles, responsibilities and authorities, as applicable (see 5.3);
>  >
>  >  4) determining how to fulfil legal requirements and other requirements (see 6.1.3);

5) establishing OH&S objectives and planning to achieve them (see 6.2);

6) determining applicable controls for outsourcing, procurement and contractors (see 8.1.4);

7) determining what needs to be monitored, measured and evaluated (see 9.1);

8) planning, establishing, implementing and maintaining an audit programme(s) (see 9.2.2);

9) ensuring continual improvement (see 10.3);

e) emphasize the **participation** of non-managerial workers in the following:

1) determining the mechanisms for their consultation and **participation**;

2) identifying hazards and assessing risks and opportunities (see 6.1.1 and 6.1.2);

3) determining actions to eliminate hazards and reduce OH&S risks (see 6.1.4);

4) determining competence requirements, training needs, training and evaluating training (see 7.2);

5) determining what needs to be communicated and how this will be done (see 7.4);

6) determining control measures and their effective implementation and use (see 8.1, 8.1.3 and 8.2);

7) investigating incidents and nonconformities and determining corrective actions (see 10.2).

NOTE 3 Emphasizing the consultation and **participation** of non-managerial workers is intended to apply to persons carrying out the work activities, but is not intended to exclude, for example, managers who are impacted by work activities or other factors in the organisation.

NOTE 4 It is recognised that the provision of training at no cost to workers and the provision of training during working hours, where possible, can remove significant barriers to worker **participation.**

## 9.3 Management review

The management review shall include consideration of:

d) information on the OH&S performance, including trends in:

   5) consultation and **participation** of workers;

## 10.2 Incident, nonconformity and corrective action

When an incident or a nonconformity occurs, the organisation shall:

b) evaluate, with the **participation** of workers (see 5.4) and the involvement of other relevant interested parties, the need for corrective action to eliminate the root cause(s) of the incident or nonconformity, in order that it does not recur or occur elsewhere, by:

## 10.3 Continual improvement

The organisation shall continually improve the suitability, adequacy and effectiveness of the OH&S management system, by:

c) promoting the **participation** of workers in implementing actions for the continual improvement of the OH&S management system;

**ISO 45001 Annex A:**

**A.5.4 Consultation and participation of workers**

The consultation and **participation** of workers, and, where they exist, workers' representatives, can be key factors of success for an OH&S management system and should be encouraged through the processes established by the organisation.

Consultation implies a two-way communication involving dialogue and exchanges. Consultation involves the timely provision of the information necessary for workers, and, where they exist, workers' representatives, to give informed feedback to be considered by the organisation before making a decision.

**Participation** enables workers to contribute to decision-making processes on OH&S performance measures and proposed changes.

Feedback on the OH&S management system is dependent upon worker **participation.** The organisation should ensure workers at all levels are encouraged to report hazardous situations, so that preventive measures can be put in place and corrective action taken.

The reception of suggestions will be more effective if workers do not fear the threat of dismissal, disciplinary action or other such reprisals when making them.

**A.6.1 Actions to address risks and opportunities**

**A.6.1.1 General**

Examples of other opportunities to improve OH&S performance:

— improving the process(es) for worker consultation and **participation;**

## C2. Safety and Health Committee and Sub-Committees

*"Mutual interests"*

### Element Description:

A health and safety committee consists of representatives from both management and workers that come together regularly to discuss and hold consultation on safety and health matters and decisions.

Health and safety committees typically:

- facilitate consultation and cooperation between management and workers

- represent and consider the views of workers

- make recommendations on safety and health matters

- conduct safety and health inspections

- review incident investigation findings

- facilitate safety promotion activities

- strengthen safety culture

The safety and health committee should safeguard the mutual interests of both the organisation and the workers and improve safety through building trust and cooperation.

Subcommittees may be formed to support the work of the main committee. Subcommittees may be tasked with:

- risk assessment/JHA
- safety promotion
- accident investigation
- housekeeping

Consultation implies a two-way communication involving dialogue and exchanges. Consultation involves the timely provision of the information necessary for workers and, where they exist, workers' representatives to give informed feedback to be considered by the organisation before making a decision.

Participation enables workers to contribute to decision-making processes on OH&S performance measures and proposed changes.

Feedback on the OH&S management system is dependent upon worker participation. The organisation should ensure workers at all levels are encouraged to report hazardous situations so that preventive measures can be put in place and corrective action taken.

The reception of suggestions will be more effective if workers do not fear the threat of dismissal, disciplinary action or other such reprisals when making them.

**Element Process Steps:**

1. establish OH&S committee charter
2. identify OH&S committee members and workers' representatives
3. train OH&S committee members

4. conduct OH&S committee meetings and inspections

5. evaluate effectiveness and improve

**ISO 45001 Requirements:**

**3.4**

**participation**

involvement in decision-making

Note 1 to entry: Participation includes engaging health and safety committees and workers' representatives, where they exist.

**4.2 Understanding the needs and expectations of workers and other interested parties**

The organisation shall determine:

b) the relevant needs and expectations (i.e. requirements) of workers and other interested parties;

**5.1 Leadership and commitment**

Top management shall demonstrate leadership and commitment with respect to the OH&S management system by:

l) ensuring the organisation establishes and implements a process(es) for consultation and participation of workers (see 5.4);

m) supporting the establishment and functioning of health and safety committees, [see 5.4 e) 1)].

### 5.4 Consultation and participation of workers

The organisation shall establish, implement and maintain a process(es) for consultation and participation of workers at all applicable levels and functions, and, where they exist, workers' representatives, in the development, planning, implementation, performance evaluation and actions for improvement of the OH&S management system.

### ISO 45001 Annex A:

### A.7.2 Competence

As appropriate, workers should receive the training required to enable them to carry out their representative functions for occupational health and safety effectively.

### A.5.4 Consultation and participation of workers

The consultation and participation of workers, and, where they exist, workers' representatives, can be key factors of success for an OH&S management system and should be encouraged through the processes established by the organisation.

## C3. Safety and Health Promotional Programme

*"Safety as a value"*

### Element Description:

A health and safety promotional programme is basically the internal marketing programme to workers for safety and health.

Promotions can be very effective in:

- improving safety and health awareness

- increasing safety and health knowledge
- encouraging employees to work safely
- cultivating safety culture

Effective strategies utilise the fastest methods to engage workers' interest and motivate workers to apply the safety and health information communicated.

OH&S Promotion includes:

- OH&S noticeboards
- OH&S displays and audiovisuals
- OH&S pamphlets, cards, and handbooks
- OH&S banners and slogans
- OH&S statistics

Promotional programmes typically include the organisation of OH&S Campaigns. OH&S Campaigns can be organised for:

- OH&S milestones achieved
- critical OH&S topic

OH&S Campaigns typically include:

- senior management involvement
- OH&S talks and speeches
- contests and quizzes
- worker participation

- promotional items
- OH&S awards, rewards, and tokens of appreciation

Off-the-job health and safety promotion may also be organised for home and family safety.

**Element Process Steps:**

1. identify critical topics for OH&S promotion
2. establish annual safety promotion programme
3. implement promotional programme
4. collect and analyse data
5. evaluate effectiveness and improve

**ISO 45001 Requirements:**

**5.1 Leadership and commitment**

Top management shall demonstrate leadership and commitment with respect to the OH&S management system by:

e) communicating the importance of effective OH&S management and of conforming to the OH&S management system requirements;

h) ensuring and **promoting** continual improvement;

j) developing, leading and **promoting a culture** in the organisation that supports the intended outcomes of the OH&S management system;

### 10.3 Continual improvement

The organisation shall continually improve the suitability, adequacy and effectiveness of the OH&S management system, by:

a) enhancing OH&S performance;

b) **promoting a culture** that supports an OH&S management system;

### ISO 45001 Annex A:

### A.7.3 Awareness

In addition to workers (especially temporary workers), contractors, visitors and any other parties should be aware of the OH&S risks to which they are exposed.

## C4. Safety and Health Awards and Recognition

*"Motivation"*

### Element Description:

Health and safety awards and recognition, when properly managed, can motivate performance in safety and health.

Awards and recognition should be given based on:

- compliance with requirements
- safety and health knowledge
- safety and health suggestions
- safety and health innovations or contributions

- observations and reporting of events
- safety and health performance (leading indicators only)
- housekeeping

Awards and recognition should not be primarily based on lagging indicators.

### Element Process Steps:

1. establish individual and group award system
2. communicate award criteria
3. implement system
4. gather feedback on award system
5. evaluate effectiveness and improve

### ISO 45001 Requirements:

### 5.1 Leadership and commitment

Top management shall demonstrate leadership and commitment with respect to the OH&S management system by:

h) ensuring and **promoting** continual improvement;

j) developing, leading and **promoting a culture** in the organisation that supports the intended outcomes of the OH&S management system;

## 10.3 Continual improvement

The organisation shall continually improve the suitability, adequacy and effectiveness of the OH&S management system, by:

a) enhancing OH&S performance;

b) **promoting a culture** that supports an OH&S management system;

c) **promoting** the participation of workers in implementing actions for the continual improvement of the OH&S management system;

### ISO 45001 Annex A:

### A.5.4 Consultation and participation of workers

The consultation and participation of workers, and, where they exist, workers' representatives, can be key factors of success for an OH&S management system and should be **encouraged** through the processes established by the organisation.

## C5. Hazard, Concern, Near Miss, and Whistle-Blower Reporting System

*"Open communication"*

### Element Description:

There should be mechanisms to allow workers to report:

- opportunities for improvement, suggestions

- H&S concerns

- hazards
- near misses
- accidents
- violations and negligence (whistle-blower reporting)

In order to learn from incidents and prevent recurrence of accidents, it is crucial that incidents and near-misses are reported. In most organisation, near-miss reporting is usually poor for a number of reasons.

Workers are reluctant to report hazards, near misses, or concerns when:

- no action will be taken
- they or their co-workers get blamed or penalised somehow
- they end up having more work to do
- they get a negative response

Few organisations have a successful reporting culture because trust needs to be built up over time; it does not happen overnight. Many organisations have punitive cultures that discourage reporting.

Strong hazard, near miss, and concern reporting is a sign of a strong safety culture.

It includes:

- reporting process or mechanism
- assigned responsibilities to respond

- action tracking

- data analysis

There should be a process to evaluate near misses reported to determine the severity of worst-case scenarios. High-potential near misses should be subject to root cause analysis to prevent recurrence of any similar incidents.

Management processes should be in place to encourage reporting. Obstacles and barriers to reporting should be identified and removed by the organisation. Goals and targets of "zero incidents or accidents" must not result in workers being discouraged from reporting any incidents. Measures should be in place to encourage workers to report any incidents or OH&S concerns.

The whistle-blower programme allows a worker to formally file a complaint in relation to a health and safety issue, concern, or violation with designated personnel.

A worker should be informed of his or her right to refuse to work or do particular work where he or she has reason to believe that this work may endanger himself or herself, or another worker.

The programme must protect workers from any reprisal. This scope of programme should cover:

- unresolved health and safety risks, actual or potential, arising from work

- obstacles and barriers to participation in OH&S

- reprisals or threat of reprisal, including disciplinary measures, reduced remuneration or terms or privileges of employment, blacklist, termination, suspension, demotion, reduction in salary, failure to hire, or any act that would deter a reasonable person from engaging in participation or whistle-blowing

Mechanism for whistle-blowing should be established and communicated to all workers and contractors. The process may also allow workers to whistle-blow anonymously. The name of the whistle-blower should be kept confidential and only released on a "need to know" basis.

A formalised process should be in place to carry out investigation of any cases of "refusal to work" and whistle-blowing. Teams or persons appointed to carry out the investigation must be able to maintain objectivity and impartiality of the investigation process.

Pending the results of the investigation, the related work should be suspended if necessary. Results of the investigation must be communicated to the worker or whistle-blower (if named) in a timely manner.

Obstacles and barriers to reporting should be identified and removed by the organisation.

**Element Process Steps:**

1. establish channels for reporting
2. promote reporting
3. take prompt corrective actions or consultation
4. give recognition for reporting
5. evaluate effectiveness and improve

## ISO 45001 Requirements:

### 5.1 Leadership and commitment

Top management shall demonstrate leadership and commitment with respect to the OH&S management system by:

j) developing, leading and promoting a culture in the organisation that supports the intended outcomes of the OH&S management system;

k) protecting workers from reprisals when **reporting incidents, hazards, risks and opportunities;**

## ISO 45001 Annex A:

### A.5.1 Leadership and commitment

An important way top management demonstrates leadership is by encouraging workers to **report incidents, hazards, risks and opportunities** and by protecting workers against reprisals, such as the threat of dismissal or disciplinary action, when they do so.

### A.5.4 Consultation and participation of workers

The organisation should ensure workers at all levels are encouraged to **report hazardous situations,** so that preventive measures can be put in place and corrective action taken.

### A.6.1 Actions to address risks and opportunities

### A.6.1.1 General

Examples of other opportunities to improve OH&S performance:

— improving the occupational health and safety culture, such as by extending competence related to occupational health and safety beyond requirements or encouraging workers to **report incidents** in a timely manner;

— improving the visibility of top management's support for the OH&S management system;

### A.7.2 Competence

Workers should have the necessary competence to remove themselves from situations of imminent and serious danger. For this purpose, it is important that workers are provided with sufficient training on hazards and risks associated with their work.

## C6. Leadership

*"Leading safety"*

**Element Description:**

Effective leadership is critical to the success of the OH&S Management System. Good safety leadership is only the beginning of a strong safety culture in organisations.

All key elements of the OH&S Management System should be championed by a senior leader.

Safety leadership is required at all levels of an organisation:

- top management
- middle management
- line supervisory

Leaders need to be involved in various elements of the system, such as:

- risk assessment
- hazard analysis
- communication
- coordination
- personal safety contacts

In an organisation with strong safety leadership, leaders:

- take on safety ownership and responsibilities
- have good knowledge of safety management
- communicate expectations
- monitor safety performance
- reward and commend good safety performance
- show genuine interest in work safety issues

Leaders must also set good safety examples wherever they go.

All senior, middle, and frontline management should be given the appropriate level of safety management training, specifically on the elements that they will be involved in, so that they can provide the required level of leadership. It is difficult to get management to support a safety programme if they do not understand the purpose, benefits, or implementation of the programme.

**Element Process Steps:**

1. identify areas where leaders can be involved
2. set safety leadership actions and programmes
3. implement leadership actions and programmes
4. evaluate effectiveness and improve

**ISO 45001 Requirements:**

**0.3 Success factors**

The implementation of an OH&S management system is a strategic and operational decision for an organisation. The success of the OH&S management system depends on leadership, commitment and participation from all levels and functions of the organisation.

The implementation and maintenance of an OH&S management system, its effectiveness and its ability to achieve its intended outcomes are dependent on a number of key factors, which can include:

a) top management leadership, commitment, responsibilities and accountability;

**5.1 Leadership and commitment**

Top management shall demonstrate leadership and commitment with respect to the OH&S management system by:

a) taking overall responsibility and accountability for the prevention of work-related injury and ill health, as well as the provision of safe and healthy workplaces and activities;

b) ensuring that the OH&S policy and related OH&S objectives are established and are compatible with the strategic direction of the organisation;

c) ensuring the integration of the OH&S management system requirements into the organisation's business processes;

d) ensuring that the resources needed to establish, implement, maintain and improve the OH&S management system are available;

e) communicating the importance of effective OH&S management and of conforming to the OH&S management system requirements;

f) ensuring that the OH&S management system achieves its intended outcome(s);

g) directing and supporting persons to contribute to the effectiveness of the OH&S management system;

h) ensuring and promoting continual improvement;

i) supporting other relevant management roles to demonstrate their leadership as it applies to their areas of responsibility;

j) developing, leading and promoting a culture in the organisation that supports the intended outcomes of the OH&S management system;

k) protecting workers from reprisals when reporting incidents, hazards, risks and opportunities;

l) ensuring the organisation establishes and implements a process(es) for consultation and participation of workers (see 5.4);

m) supporting the establishment and functioning of health and safety committees, [see 5.4 e) 1)].

**ISO 45001 Annex A:**

**A.5.1 Leadership and commitment**

Leadership and commitment from the organisation's top management, including awareness, responsiveness, active support and feedback, are critical for the success of the OH&S management system and achievement of its intended outcomes; therefore, top management has specific responsibilities for which they need to be personally involved or which they need to direct.

A culture that supports an organisation's OH&S management system is largely determined by top management and is the product of individual and group values, attitudes, managerial practises, perceptions, competencies and patterns of activities that determine the commitment to, and the style and proficiency of, its OH&S management system. It is characterized by, but not limited to, active participation of workers, cooperation and communications founded on mutual trust, shared perceptions of the importance of the OH&S management system by active involvement in detection of OH&S opportunities and confidence in the effectiveness of preventive and protective measures. An important way top management demonstrates leadership is by encouraging workers to report incidents, hazards, risks and opportunities and by protecting workers against reprisals, such as the threat of dismissal or disciplinary action, when they do so.

## C7. Safety Culture

*"Values and norms"*

### Element Description:

Safety culture is the product of individual and group values, attitudes, perceptions, competencies, and patterns of behaviour that determine their commitment to safety. It's also the style and proficiency of an organisation's health and safety management.

A safety culture programme seeks to progressively strengthen the safety culture over a period of time. Because the safety culture of an organisation cannot be changed overnight, time is needed for the organisation to grow trust, values, involvement, and cooperation.

Safety culture is unique to each organisation and has to be suitable to the context of the organisation, risk, and business environment.

Common characteristics of a strong safety culture:

- trust
- leadership
- open communication
- open reporting of incidents and safety concerns
- high level of safety knowledge
- teamwork
- involvement
- care and concern

- fairness
- refrain from a "blame" environment
- continual safety improvement
- constant state of awareness and vigilance
- willingness to stop work to address safety concerns

Otherwise aspects of safety culture that may vary among companies:

- rites and rituals
- stories and history
- communication climate
- shared assumptions
- openness to innovation and change

Attributes of safety culture include:

- leadership
- employee involvement
- line ownership
- teamwork and cooperation
- communication
- reporting and feedback
- fair, just culture and trust

- risk awareness and attitudes
- learning and continual improvement
- discipline

Employees should be constantly vigilant about safety to the point of chronic unease to avoid any complacency and risk habituation.

A safety culture programme may include the following:

- identification of core values
- safety goals
- integration of safety into vision
- safety initiatives
- leadership activities
- employee involvement
- communications
- activities

Safety culture should be measured periodically. Actions needed to improve safety culture should be identified.

Safety climate surveys should be carried out periodically.

An organisation that aspires to develop a strong safety culture should also implement an off-the-job safety programme because safety as a value cannot be confined to just the workplace. Safety should also be practised outside the workplace and in the home if it is to become part of workers' value systems.

### Element Process Steps:

1. establish safety culture programme
2. conduct leadership training
3. assess safety culture
4. identify actions to improve culture
5. implement culture-building programme
6. evaluate effectiveness and improve

### ISO 45001 Requirements:

**0.3 Success factors**

The implementation of an OH&S management system is a strategic and operational decision for an organisation. The success of the OH&S management system depends on leadership, commitment and participation from all levels and functions of the organisation.

The implementation and maintenance of an OH&S management system, its effectiveness and its ability to achieve its intended outcomes are dependent on a number of key factors, which can include:

a) top management leadership, commitment, responsibilities and accountability;

b) top management developing, leading and promoting a culture in the organisation that supports the intended outcomes of the OH&S management system;

## 5.1 Leadership and commitment

Top management shall demonstrate leadership and commitment with respect to the OH&S management system by:

j) developing, leading and promoting a culture in the organisation that supports the intended outcomes of the OH&S management system;

## 10.3 Continual improvement

The organisation shall continually improve the suitability, adequacy and effectiveness of the OH&S management system, by:

b) promoting a culture that supports an OH&S management system;

## ISO 45001 Annex A:

### A.5.1 Leadership and commitment

A culture that supports an organisation's OH&S management system is largely determined by top management and is the product of individual and group values, attitudes, managerial practises, perceptions, competencies and patterns of activities that determine the commitment to, and the style and proficiency of, its OH&S management system. It is characterized by, but not limited to, active participation of workers, cooperation and communications founded on mutual trust, shared perceptions of the importance of the OH&S management system by active involvement in detection of OH&S opportunities and confidence in the effectiveness of preventive and protective measures. An important way top management demonstrates leadership is by encouraging workers to report incidents, hazards, risks and opportunities and by protecting workers against reprisals, such as the threat of dismissal or disciplinary action, when they do so.

### A.6.1 Actions to address risks and opportunities

### A.6.1.1 General

Examples of other opportunities to improve OH&S performance:

— improving the occupational health and safety culture, such as by extending competence related to occupational health and safety beyond requirements or encouraging workers to report incidents in a timely manner;

### D1. PPE Programme

*"Hard hats and safety shoes"*

**Element Description:**

Personal protective equipment (PPE) is personal equipment worn to protect from injury or minimise exposure to physical, chemical, biological, or ergonomic hazards. The purpose of PPE is to reduce worker's exposure to hazards when engineering controls and administrative controls are not feasible or effective to reduce these risks to acceptable levels. PPE guidelines can be effective in controlling the residual risk if they are effectively implemented and followed.

A PPE programme typically includes the following elements:

- identification of PPE needs
- selection of PPE
- training on PPE usage

- PPE matrix or requirements
- communication of PPE requirements
- fitting and sizing of PPE
- maintenance and inspection of PPE
- PPE enforcement and compliance checks
- PPE records

**Element Process Steps:**

1. identify PPE needs
2. establish PPE matrix and requirements
3. select PPE in consultation with workers
4. issue PPE to workers
5. monitor and enforce PPE use
6. evaluate effectiveness and improve

**ISO 45001 Requirements:**

**8.1.2 Eliminating hazards and reducing OH&S risks**

The organisation shall establish, implement and maintain a process(es) for the elimination of hazards and reduction of OH&S risks using the following hierarchy of controls:

e) use adequate personal protective equipment.

NOTE In many countries, legal requirements and other requirements include the requirement that personal protective equipment (PPE) is provided at no cost to workers.

## ISO 45001 Annex A:

### A.8.1.2 Eliminating hazards and reducing OH&S risks

The hierarchy of controls is intended to provide a systematic approach to enhance occupational health and safety, eliminate hazards, and reduce or control OH&S risks. Each control is considered less effective than the one before it. It is usual to combine several controls in order to succeed in reducing the OH&S risks to a level that is as low as reasonably practicable.

The following examples are given to illustrate measures that can be implemented at each level.

e) **Personal protective equipment (PPE):** providing adequate PPE, including clothing and instructions for PPE utilization and maintenance (e.g. safety shoes, safety glasses, hearing protection, gloves).

### A.10.2 Incident, nonconformity and corrective action

Examples of incidents, nonconformities and corrective actions can include, but are not limited to:

c) corrective actions (as indicated by the hierarchy of controls; see 8.1.2): eliminating hazards; substituting with less hazardous materials; redesigning or modifying equipment or tools; developing procedures; improving the competence of affected workers; changing the frequency of use; using **personal protective equipment**.

## D2. High-Risk Operations Control Programme

*"Critical controls"*

### Element Description:

The programme is to identify relatively higher risk operations so that focus can be maintained on the controls to manage the risk from these activities.

High-risk operations are activities that can result in major losses if they are not carried out properly.

Typical high-risk operations are:

- lifting
- entry into confined spaces
- high-pressure testing
- radiography
- working at height
- hot work
- line breaking
- excavation
- transportation

Planning and controls for high-risk operations needs to be implemented in advance. Controls may link to other relevant elements such as:

- permit-to-work

- safe work practices
- coordination meetings
- risk assessment and hazard analysis
- PPE

A hierarchy of control should be followed where feasible.

Other typical controls may include, where relevant:

- qualified personnel
- proper planning
- identification of legal requirements
- procedures, instructions, or method statements
- pre-activity checks and equipment inspections
- verification of control implementation
- validation of engineering design
- hazard communication
- control and coordination plan
- risk monitoring

**Element Process Steps:**

1. identify all high-risk operations
2. analyse high-risk operations (e.g., bowtie)
3. identify barriers and controls

4. formalise requirements into the system

5. implement controls

6. evaluate effectiveness and improve

**ISO 45001 Requirements:**

**6.1 Actions to address risks and opportunities**

**6.1.1 General**

When planning for the OH&S management system, the organisation shall consider the issues referred to in 4.1 (context), the requirements referred to in 4.2 (interested parties) and 4.3 (the scope of its OH&S management system) and determine the **risks** and opportunities that need to be addressed to:

a) give assurance that the OH&S management system can achieve its intended outcome(s);

b) prevent, or reduce, undesired effects;

c) achieve continual improvement.

When determining the **risks** and opportunities for the OH&S management system and its intended outcomes that need to be addressed, the organisation shall take into account:

— hazards (see 6.1.2.1);

— OH&S **risks** and other risks (see 6.1.2.2);

— OH&S opportunities and other opportunities (see 6.1.2.3);

— legal requirements and other requirements (see 6.1.3).

The organisation, in its planning process(es), shall determine and assess the **risks** and opportunities that are relevant to the intended outcomes of the OH&S management system associated with changes in the organisation, its processes or the OH&S management system.

### 6.1.2.3 Assessment of OH&S opportunities and other opportunities for the OH&S

### management system

The organisation shall establish, implement and maintain a process(es) to assess:

a) OH&S opportunities to enhance OH&S performance, while taking into account planned changes to the organisation, its policies, its processes or its activities and:

1) opportunities to adapt work, work organisation and work environment to workers;

2) opportunities to eliminate hazards and reduce OH&S **risks**;

### 6.1.4 Planning action

The organisation shall plan:

a) actions to:

1) address these **risks** and opportunities (see 6.1.2.2 and 6.1.2.3);

### 8.1.2 Eliminating hazards and reducing OH&S risks

The organisation shall establish, implement and maintain a process(es) for the elimination of hazards and reduction of OH&S **risks** using the following hierarchy of controls:

a) eliminate the hazard;

b) substitute with less hazardous processes, operations, materials or equipment;

c) use engineering controls and reorganisation of work;

d) use administrative controls, including training;

e) use adequate personal protective equipment.

**ISO 45001 Annex A:**

**A.6.1.4 Planning action**

When the assessment of OH&S **risks** and other risks has identified the need for controls, the planning activity determines how these are implemented in operation (see Clause 8); for example, determining whether to incorporate these controls into work instructions or into actions to improve competence. Other controls can take the form of measuring or monitoring (see Clause 9).

Actions to address **risks** and opportunities should also be considered under the management of change (see 8.1.3) to ensure there are no resulting unintended consequences.

## D3. Hazardous Chemicals Control Programme

*"Chemical safety"*

**Element Description:**

Hazardous chemicals, if handled or stored improperly, may result in accidents. A comprehensive programme should be in place to ensure that hazardous chemicals are properly managed. The programme should include, where relevant:

- identification of legal requirements and required licenses
- training of relevant personnel
- risk assessment and hazard analysis
- management of safety information such as safety data sheets
- labelling and warning signs
- handling and transportation
- proper storage facility
- inventory control
- safe work practices and/or operation procedure
- emergency preparedness (e.g., first aid, antidotes, spill control)
- personal protective equipment
- hygiene monitoring and/or medical surveillance
- inspection and audit
- disposal of hazardous wastes

There should be a process to evaluate new chemicals for their hazards and precautions before they can be purchased.

All control measures should be implemented before hazardous chemicals can be used.

## Element Process Steps:

1. establish chemical inventory
2. identify all hazardous substances
3. establish control measures
4. communicate hazard
5. conduct monitoring
6. evaluate effectiveness and improve

## ISO 45001 Requirements:

### 6.1.2.1 Hazard identification

The organisation shall establish, implement and maintain a process(es) for hazard identification that is ongoing and proactive. The process(es) shall take into account, but not be limited to:

   a) how work is organised, social factors (including workload, work hours, victimization, harassment and bullying), leadership and the culture in the organisation;

   b) routine and non-routine activities and situations, including hazards arising from:

1) infrastructure, equipment, **materials, substances** and the physical conditions of the workplace;

### 8.1.2 Eliminating hazards and reducing OH&S risks

The organisation shall establish, implement and maintain a process(es) for the elimination of hazards and reduction of OH&S risks using the following hierarchy of controls:

a) eliminate the hazard;

b) substitute with less **hazardous** processes, operations, **materials** or equipment;

## ISO 45001 Annex A:

### A.6.1.2.1 Hazard identification

Hazard identification helps the organisation recognise and understand the hazards in the workplace and to workers, in order to assess, prioritize and eliminate hazards or reduce OH&S risks.

Hazards can be physical, **chemical,** biological, psychosocial, mechanical, electrical or based on movement and energy.

### A.8.1.2 Eliminating hazards and reducing OH&S risks

The following examples are given to illustrate measures that can be implemented at each level.

a) Elimination: removing the hazard; stopping using **hazardous chemicals;** applying ergonomics approaches when planning new workplaces; eliminating monotonous work or work that causes negative stress; removing fork-lift trucks from an area.

b) Substitution: replacing the **hazardous** with less hazardous; changing to answering customer complaints with online guidance; combating OH&S risks at source; adapting to technical progress (e.g. replacing solvent-based paint by water-based paint; changing slippery floor material; lowering voltage requirements for equipment).

### A.8.1.4 Procurement

### A.8.1.4.1 General

The procurement process(es) should be used to determine, assess and eliminate hazards, and to reduce OH&S risks associated with, for example, products, **hazardous materials or substances,** raw materials, equipment, or services before their introduction into the workplace.

### A.9.1 Monitoring, measurement, analysis and performance evaluation

### A.9.1.1 General

Measurement generally involves the assignment of numbers to objects or events. It is the basis for quantitative data and is generally associated with the performance evaluation of safety programmes and health surveillance. Examples include the use of calibrated or verified equipment to measure exposure to a hazardous substance or the calculation of the safe distance from a hazard.

### D4. Health Control Programmes

*"Healthy"*

### Element Description:

Health hazards may be classified into the following categories:

- physical hazards (e.g., noise)

- chemical hazards (e.g., carcinogens)

- ergonomic and psychosocial hazards (e.g., stress)

- biological hazards (e.g., viruses)

Common health hazards may include:

- hazardous chemicals
- respiratory hazards
- noise hazards

Health control programmes are needed for the relevant health hazards to which workers are exposed.

Health control programmes may include:

- engineering control measures
- hazard identification and evaluation
- PPE

### Element Process Steps:

1. identification of health hazards
2. conduct health risk assessments
3. evaluation of health hazards
4. establish control programmes
5. implement programmes
6. evaluate effectiveness and improve

## ISO 45001 Requirements:

### 6.1.2.1 Hazard identification

The organisation shall establish, implement and maintain a process(es) for hazard identification that is ongoing and proactive.

### 6.1.4 Planning action

The organisation shall plan:

a) actions to:

1) address these risks and opportunities (see 6.1.2.2 and 6.1.2.3);

2) address legal requirements and other requirements (see 6.1.3);

### 8.1.2 Eliminating hazards and reducing OH&S risks

The organisation shall establish, implement and maintain a process(es) for the elimination of hazards and reduction of OH&S risks using the following hierarchy of controls:

a) eliminate the hazard;

b) substitute with less hazardous processes, operations, materials or equipment;

c) use engineering controls and reorganisation of work;

d) use administrative controls, including training;

e) use adequate personal protective equipment.

## D5. Hygiene Monitoring and Medical Surveillance Programme

*"Check conditions"*

**Element Description:**

Hygiene monitoring is a process of evaluating workers' exposure to biological, chemical, and physical health hazards. Monitoring can be qualitative or quantitative.

Hygiene monitoring programme should identify the substances that require monitoring. There may be legal requirements for hygiene monitoring.

Hygiene monitoring programme includes:

- identification of hazardous substance
- exposure monitoring
- evaluation of exposure controls
- evaluation of monitoring results

Qualified personnel such as an industrial hygienist may be required for workplaces with significant exposures.

Medical surveillance is the systematic assessment of workers exposed to occupational health hazards. Medical surveillance describes activities that target health events or a change in a biologic function of exposed workers. A medical surveillance programme involves recurrent examinations and data analysis over time.

Medical surveillance programme includes:

- identification of health hazard

- identification of legal requirements
- selection of personnel for surveillance
- medical examinations
- interpretation and communication of test results

**Element Process Steps:**

1. identify exposures requiring monitoring
2. establish hygiene monitoring and medical surveillance programme
3. evaluate monitoring results
4. implement exposure control measures
5. evaluate effectiveness and improve

**ISO 45001 Requirements:**

**9.1 Monitoring, measurement, analysis and performance evaluation**

**9.1.1 General**

The organisation shall establish, implement and maintain a process(es) for **monitoring**, measurement, analysis and performance evaluation.

The organisation shall determine:

a) what needs to be monitored and measured, including:

1) the extent to which legal requirements and other requirements are fulfilled;

2) its activities and operations related to **identified hazards**, risks and opportunities;

3) progress towards achievement of the organisation's OH&S objectives;

4) effectiveness of operational and other controls;

b) the methods for monitoring, measurement, analysis and performance evaluation, as applicable, to ensure valid results;

c) the criteria against which the organisation will evaluate its OH&S performance;

d) when the monitoring and measuring shall be performed;

e) when the results from monitoring and measurement shall be analysed, evaluated and communicated.

The organisation shall ensure that monitoring and measuring equipment is calibrated or verified as applicable, and is used and maintained as appropriate.

NOTE There can be legal requirements or other requirements (e.g. national or international standards) concerning the calibration or verification of monitoring and measuring equipment.

The organisation shall retain appropriate documented information:

— as evidence of the results of monitoring, measurement, analysis and performance evaluation;

— on the maintenance, calibration or verification of measuring equipment.

**ISO 45001 Annex A:**

**A.8.1.2 Eliminating hazards and reducing OH&S risks**

The following examples are given to illustrate measures that can be implemented at each level.

d) Administrative controls including training: conducting periodic safety equipment inspections; conducting training to prevent bullying and harassment; managing health and safety coordination with subcontractors' activities; conducting induction training; administrating forklift driving licences; providing instructions on how to report incidents, nonconformities and victimization without fear of retribution; changing the work patterns (e.g. shifts) of workers; managing a **health or medical surveillance programme** for workers who have been identified as at risk (e.g. related to hearing, hand-arm vibration, respiratory disorders, skin disorders or exposure); giving appropriate instructions to workers (e.g. entry control processes).

**A.9.1 Monitoring, measurement, analysis and performance evaluation**

**A.9.1.1 General**

In order to achieve the intended outcomes of the OH&S management system, the processes should be monitored, measured and analysed.

a) Examples of what could be monitored and measured can include, but are not limited to:

1) occupational health complaints, **health of workers (through surveillance) and work environment;**

2) work-related incidents, injuries and ill health, and complaints, including trends;

Measurement generally involves the assignment of numbers to objects or events. It is the basis for quantitative data and is generally associated with the performance evaluation of safety programmes and **health surveillance**. Examples include the use of calibrated or verified equipment to measure exposure to a hazardous substance or the calculation of the safe distance from a hazard.

## D6. Ergonomics and Fatigue Management Programme

*"Human factors"*

### Element Description:

This programme is about identifying, analysing, and controlling ergonomic risk factors and fitting workplace conditions and job demands to the capabilities of the workers.

- identification of risk factors
- ergonomic team training
- ergonomic assessment
- ergonomic interventions

Fatigue is a state of impairment that can include physical and/or mental elements associated with lower alertness and reduced performance. Fatigue can lead to incidents because workers are not alert and are less able to respond to changing circumstances. Fatigue can also lead to long-term health problems.

The purpose of a fatigue management programme is to assess the risk of fatigue and mitigate the effects of fatigue through a range of control measures.

A fatigue management programme may include:

- fatigue risk assessment
- scheduling and rostering
- fitness for duty
- fatigue knowledge and awareness
- consultation with workers
- implementation of controls
- review programme effectiveness

**Element Process Steps:**

1. establish ergonomic programme
2. train ergonomic assessors
3. identify activities with ergonomic issues
4. conduct ergonomic assessments
5. implement changes
6. evaluate effectiveness and improve

**ISO 45001 Requirements:**

**6.1.2.1 Hazard identification**

The organisation shall establish, implement and maintain a process(es) for hazard identification that is ongoing and proactive. The process(es) shall take into account, but not be limited to:

b) routine and non-routine activities and situations, including hazards arising from:

3) human factors;

**8.1 Operational planning and control**

**8.1.1 General**

The organisation shall plan, implement, control and maintain the processes needed to meet requirements of the OH&S management system, and to implement the actions determined in Clause 6, by:

d) adapting work to workers.

**ISO 45001 Annex A:**

**A.6.1 Actions to address risks and opportunities**

**A.6.1.1 General**

OH&S opportunities address the identification of hazards, how they are communicated, and the analysis and mitigation of known hazards. Other opportunities address system improvement strategies.

Examples of opportunities to improve OH&S performance:

c) improving OH&S performance by alleviating **monotonous work** or work at a potentially hazardous pre-determined work rate;

f) **ergonomic** and other injury prevention-related assessments.

**A.6.1.2 Hazard identification and assessment of risks and opportunities**

**A.6.1.2.1 Hazard identification**

The organisation's hazard identification process(es) should consider:

b) human factors:

1) relate to human capabilities, limitations and other characteristics;

2) information should be applied to tools, machines, systems, activities and environment for safe, comfortable human use;

3) should address three aspects: the activity, the worker and the organisation, and how these interact with and impact on occupational health and safety;

**A.8.1 Operational planning and control**

**A.8.1.1 General**

Examples of operational control of the processes include:

g) adapting work to workers; for example, by:

1) defining, or redefining, how the work is organised;

2) the induction of new workers;

3) defining, or redefining, processes and working environments;

4) using **ergonomic** approaches when designing new, or modifying, workplaces, equipment, etc.

### A.8.1.2 Eliminating hazards and reducing OH&S risks

The hierarchy of controls is intended to provide a systematic approach to enhance occupational health and safety, eliminate hazards, and reduce or control OH&S risks. Each control is considered less effective than the one before it. It is usual to combine several controls in order to succeed in reducing the OH&S risks to a level that is as low as reasonably practicable.

The following examples are given to illustrate measures that can be implemented at each level.

a) Elimination: removing the hazard; stopping using hazardous chemicals; applying **ergonomics** approaches when planning new workplaces; eliminating monotonous work or work that causes negative stress; removing fork-lift trucks from an area.

b) Engineering controls, reorganisation of work, or both: isolating people from hazard; implementing collective protective measures (e.g. isolation, machine guarding, ventilation systems); addressing mechanical handling; reducing noise; protecting against falls from height by using guard rails; reorganising work to avoid people working alone, **unhealthy work hours and workload,** or to prevent victimization.

c) Administrative controls including training: conducting periodic safety equipment inspections; conducting training to prevent bullying and harassment; managing health and safety coordination with subcontractors' activities; conducting induction training; administrating forklift driving licences; providing instructions on how to report incidents, nonconformities and victimization without fear of retribution; **changing the work patterns (e.g. shifts) of workers;** managing a health or medical surveillance programme for workers who have been identified as at risk (e.g. related to hearing, hand-arm vibration, respiratory disorders, skin disorders or exposure); giving appropriate instructions to workers (e.g. entry control processes).

### A.10.2 Incident, nonconformity and corrective action

When determining the root cause of an incident or nonconformity, the organisation should use methods appropriate to the nature of the incident or nonconformity being analysed. The focus of root cause analysis is prevention. This analysis can identify multiple contributory failures, including factors related to communication, competence, **fatigue,** equipment or procedures.

### D7. Individual Risk Factors and Management

*"The individual"*

### Element Description:

Working condition may have already been well-planned and controlled such that the risk to a normal worker is minimal and safe. However, individual workers may suffer from increased risks due to other factors:

- personal characteristics (concentration ability, personality, etc.)

- personal health issues

- lone worker (lack of supervision, support, first-aid response, facilities, etc.)

- off-site worker (lack of support, security and facilities)

- young worker (knowledge, physical and behavioural issues)

- new worker (competence and awareness)

- transferred worker (unfamiliarity)

- worker back to work after long absence (unaware of changes, forgotten, etc.)

- worker coming from another climate (acclimatisation)

- worker coming from foreign country or culture (safety culture, lost in translation, etc.)

- worker coming from different social background (safety culture, teamwork and cooperation)

- worker with mental condition

- worker with health or medical condition

- worker with physical disability (e.g., hearing, colour-blind)

- temporary workers (e.g., ad-hoc workers, self-employed contractors)

There are many differences between people that have the potential to affect performance, and some individuals may be more likely to make errors than others. If necessary, workers performing critical work should be assessed (e.g., use of psychometric tests) on their concentration ability and personality.

Individuals who do not respond to coaching and counselling should be identified and may need to be removed from higher risk duties or reassigned to other jobs.

In the event of an incident, individual workers may require personalised return-to-work process and case management. A case management programme should be available to help affected individual workers return to work. Case management is a collaborative process of assessment, planning, facilitation, care coordination, evaluation, and advocacy for options and services to meet an individual's health needs through communication and available resources to promote patient safety, quality of care, and cost-effective outcomes.

### Element Process Steps:

1. establish individual risk factor programme
2. train supervisors to recognise risk factors
3. identify affected workers
4. identify control measures for applicable risk factors
5. implement controls as required
6. review each case
7. evaluate effectiveness and improve

### ISO 45001 Requirements:

### 6.1.2.1 Hazard identification

The organisation shall establish, implement and maintain a process(es) for hazard identification that is ongoing and proactive. The process(es) shall take into account, but not be limited to:

a) how work is organised, **social factors** (including workload, work hours, victimization, harassment and bullying), leadership and the culture in the organisation;

b) routine and non-routine activities and situations, including hazards arising from:

3) **human factors;**

e) people, including consideration of:

3) workers at a location not under the direct control of the organisation;

**7.2 Competence**

The organisation shall:

a) determine the necessary competence of workers that affects or can affect its OH&S performance;

NOTE Applicable actions can include, for example, the provision of training to, the mentoring of, or the reassignment of currently employed persons, or the hiring or contracting of competent persons.

**ISO 45001 Annex A:**

**A.6.1.2.1 Hazard identification**

The organisation's hazard identification process(es) should consider:

e) **people**:

2) workers at a location not under the direct control of the organisation, such as **mobile workers or workers who travel**

to perform work-related activities at another location (e.g. postal workers, bus drivers, service personnel travelling to and working at a customer's site);

3) home-based workers, or those who **work alone;**

### A.7.2 Competence

The competence of workers should include the knowledge and skills needed to appropriately identify the hazards and deal with the OH&S risks associated with their work and workplace.

In determining the competence for each role, the organisation should take into account things such as:

j) **individual** capabilities, including experience, language skills, literacy and diversity;

### A.7.3 Awareness

In addition to workers (especially **temporary workers**), contractors, visitors and any other parties should be aware of the OH&S risks to which they are exposed.

## E1. Tools, Equipment and Critical Parts Inspections

*"Check equipment"*

### Element Description:

Over time, tools and equipment may wear out or be damaged and become unsafe for use. Periodic inspections should be carried out for tools and equipment to ensure that they are safe for use.

There should be an identification system to record and show the inspection status of critical equipment. Such equipment may include:

- scaffolds
- ladders
- electrical tools

Equipment that is unsafe for use should also be identified.

Statutory inspection requirements for relevant equipment should be identified. They typically include:

- pressure vessels
- lifting appliance
- lifting platforms

Special equipment critical to OH&S should be inspected. They may include:

- fire extinguishers
- first aid kit
- SCBA
- fall arrest system

Critical parts are parts or components whose failure is most likely to result in a serious accident. There should be an inspection programme for critical parts. Typical critical parts are:

- safety relief valves
- instruments and sensors

- safety interlocks
- brakes

Pre-use inspection is a special subset of critical parts inspection that should be carried out prior to starting an equipment.

A defective equipment, when turned on, may result in accident. Pre-start (or pre-use) inspections are carried out to identify any problems related to any equipment or condition prior to commencement of operations or at the start of a shift. Typically, a checklist is used to record the inspection and to ensure that all relevant items have been inspected. Pre-start inspections are usually carried out for:

- cranes
- forklifts
- motor vehicles
- Material-handling equipment

Such checks may be carried out by maintenance or operations personnel.

**Element Process Steps:**

1. establish inventory of all tools, equipment, and machinery
2. identify critical tools, equipment, and machinery
3. identify critical parts of tools, equipment, and machinery
4. establish inspection requirements

5. develop equipment inspection checklists
6. evaluate effectiveness and improve

**ISO 45001 Requirement:**

Nil.

**ISO 45001 Annex A:**

**A.6.1 Actions to address risks and opportunities**

**A.6.1.1 General**

Examples of opportunities to improve OH&S performance:

a) **inspection** and auditing functions;

**A.8.1.1 General**

Examples of operational control of the processes include:

c) establishing preventive or predictive maintenance and **inspection** programmes;

**A.8.1.2 Eliminating hazards and reducing OH&S risks**

The following examples are given to illustrate measures that can be implemented at each level.

d) Administrative controls including training: **conducting periodic safety equipment inspections**; conducting training to prevent bullying and harassment; managing health and safety coordination with subcontractors' activities; conducting induction training; administrating forklift driving licences; providing instructions on how to report incidents, nonconformities and victimization without fear of retribution; changing the work patterns (e.g. shifts) of workers; managing a health or medical surveillance programme for workers who have been identified as at risk (e.g. related to hearing, hand-arm vibration, respiratory disorders, skin disorders or exposure); giving appropriate instructions to workers (e.g. entry control processes).

## E2. Maintenance System

*"Care for equipment"*

### Element Description:

Regular maintenance is needed to keep equipment and machines safe, reliable, and in good operating condition. Maintenance is the work of keeping something in proper condition, care, or upkeep to prevent it from breaking down.

Equipment with inadequate maintenance may break down and result in accidents.

The objective of maintenance system in safety is to ensure that plant, machinery, and equipment operate smoothly and do not present any hazard due to lack of repair or maintenance.

Maintenance Programme should be established and should include all relevant equipment.

Maintenance system should include as needed:

- breakdown maintenance
- preventive maintenance
- predictive maintenance

A system for managing maintenance work should be established. It is common for organisations to use a maintenance software to manage maintenance work.

**Element Process Steps:**

1. identify maintenance needs
2. establish maintenance philosophy
3. establish maintenance programme
4. implement maintenance programme
5. evaluate effectiveness and improve

**ISO 45001 Requirements:**

**9.1 Monitoring, measurement, analysis and performance evaluation**

**9.1.1 General**

The organisation shall ensure that monitoring and measuring equipment is calibrated or verified as applicable, and is used and **maintained** as appropriate.

The organisation shall retain appropriate documented information:

— as evidence of the results of monitoring, measurement, analysis and performance evaluation;

— on the **maintenance**, calibration or verification of measuring equipment.

## ISO 45001 Annex A:

### A.8.1 Operational planning and control

### A.8.1.1 General

Operational planning and control of the processes need to be established and implemented as necessary to enhance occupational health and safety, by eliminating hazards or, if not practicable, by reducing the OH&S risks to levels as low as reasonably practicable for operational areas and activities.

Examples of operational control of the processes include:

c) establishing preventive or predictive **maintenance** and inspection programmes;

### A.8.1.4 Procurement

### A.8.1.4.1 General

The organisation should verify that **equipment**, installations and materials are safe for use by workers by ensuring:

a) equipment is delivered according to specification and is tested to ensure it works as intended;

### A.8.1.4.2 Contractors

The need for coordination recognises that some contractors (i.e. external providers) possess specialised knowledge, skills, methods and means.

Examples of contractor activities and operations include **maintenance**, construction, operations, security, cleaning and a number of other functions.

## E3 Machine Guarding and Automation Safety Programme

*"Protecting people from machines"*

### Element Description:

Machine guarding is a common example of engineering control.

All machines and equipment generating mechanical movement should be properly guarded to prevent accidents. Machine guarding devices should:

- protect operator from hazardous moving parts
- protect workers from being struck by flying parts of material

Machine hazards may include nip points, rotating parts, moving parts, flying chips/material, and sparks.

Machine guarding may be:

- fixed
- adjustable or
- self-adjusting

Other forms of safety devices may also be considered:

- safety trip control
- two-hand control

Standards of machine guarding to be followed should be identified.

Inspections of machines and machine guarding should be carried out at defined intervals.

Automated machines introduce hazards at the point of operation and also where components transmit energy to the machine and where moving feed mechanisms and auxiliary parts are found.

A robot is an automatically controlled, reprogrammable, multipurpose, manipulator programmable in three or more axes, which may be either fixed in place or mobile for use in industrial automation applications. Robots share many of the safety issues with industrial automated machines. In addition, robots operate on more axes and with greater freedom to determine their work envelope. Robots has greater risk than automated machines because of the uncertainty of

- the speed,
- predictability of movement, and
- hazard zones.

Hazards when using robotics can come from:

- errors during use
- ejection of materials
- trapping points
- failures and malfunctions
- human entry into robot operating areas

A robot safety programme may include:

- enclosing the robot
- limits to robot motion
- use of key, plug, or actuating device

- restricting access, barriers, and perimeter safeguarding
- automatic shutdown and disabled automatic restart

All machines and robots should be provided with an emergency stop function.

As robots perform more advanced tasks, proper and accurate programming of robots is essential for safety. The design of robot work cells should include a number of safeguarding systems aimed at keeping workers at a safe distance.

Requirements for robot safety should be documented.

References:

- ISO 10218-1 Robots and robotic devices—Safety requirements for industrial robots—Part 1: Robots

**Element Process Steps:**

1. establish inventory of machines and robots
2. identify hazards requiring guarding
3. identify guarding and protection for each machine
4. implement guarding programme
5. evaluate effectiveness and improve

**ISO 45001 Requirements:**

**8.1.2 Eliminating hazards and reducing OH&S risks**

The organisation shall establish, implement and maintain a process(es) for the elimination of hazards and reduction of OH&S risks using the following hierarchy of controls:

a) eliminate the hazard;

b) substitute with less hazardous processes, operations, materials or equipment;

c) use **engineering controls** and reorganisation of work;

### ISO 45001 Annex A:

### A.6.1 Actions to address risks and opportunities

### A.6.1.1 General

Examples of other opportunities to improve OH&S performance:

— integrating occupational health and safety requirements at the earliest stage in the life cycle of facilities, equipment or process planning for facilities relocation, process re-design or replacement of **machinery** and plant;

### A.6.1.2.1 Hazard identification

Hazards can be physical, chemical, biological, psychosocial, **mechanical, electrical** or based on movement and **energy**.

The organisation's hazard identification process(es) should consider:

b) human factors:

2) information should be applied to **tools, machines**, systems, activities and environment for safe, comfortable human use;

d) potential emergency situations:

1) unplanned or unscheduled situations that require an immediate response (e.g. a **machine** catching fire in the

workplace, or a natural disaster in the vicinity of the workplace or at another location where workers are performing work-related activities);

### A.6.2.1 OH&S objectives

c) operational objectives can be set at the activity level (e.g. the enclosure of individual **machines** to reduce noise).

### A.8.1 Operational planning and control

### A.8.1.1 General

Operational planning and control of the processes need to be established and implemented as necessary to enhance occupational health and safety, by eliminating hazards or, if not practicable, by reducing the OH&S risks to levels as low as reasonably practicable for operational areas and activities.

Examples of operational control of the processes include:

a) the use of procedures and systems of work;

b) ensuring the competence of workers;

c) establishing preventive or predictive maintenance and inspection programmes;

d) specifications for the procurement of goods and services;

e) application of legal requirements and other requirements, or manufacturers' instructions for equipment;

f) **engineering** and administrative controls;

## A.8.1.2 Eliminating hazards and reducing OH&S risks

The hierarchy of controls is intended to provide a systematic approach to enhance occupational health and safety, eliminate hazards, and reduce or control OH&S risks. Each control is considered less effective than the one before it. It is usual to combine several controls in order to succeed in reducing the OH&S risks to a level that is as low as reasonably practicable.

The following examples are given to illustrate measures that can be implemented at each level.

c) Engineering controls, reorganisation of work, or both: isolating people from hazard; implementing collective protective measures (e.g. isolation, **machine guarding**, ventilation systems); addressing mechanical handling; reducing noise; protecting against falls from height by using guard rails; reorganising work to avoid people working alone, unhealthy work hours and workload, or to prevent victimization.

## A.8.1.4 Procurement

### A.8.1.4.1 General

The organisation should verify that **equipment**, installations and materials are safe for use by workers by ensuring:

a) equipment is delivered according to specification and is tested to ensure it works as intended;

## A.10.2 Incident, nonconformity and corrective action

Examples of incidents, nonconformities and corrective actions can include, but are not limited to:

b) nonconformities: **protective equipment** not functioning properly; failure to fulfil legal requirements and other requirements; prescribed procedures not being followed;

### E4. Energy Controls and Lockout Tag-Out Programme

*"Control energy"*

**Element Description:**

Energy sources including mechanical, hydraulic, electrical, pneumatic, chemical, thermal, and more in machines and equipment can cause injuries to workers. During the servicing and maintenance of machines and equipment, inadvertent release of energy or materials can result in serious injury or death to workers.

Workers servicing or maintaining machines or equipment may be seriously injured or killed if there is unexpected start-up or release of stored energy or materials.

All workers servicing and maintaining machines and equipment must be trained to recognise hazardous energy sources in the workplace, the type and magnitude of energy found in the workplace, and the means and methods of isolating and/or controlling the energy.

Devices and procedures for the control of energy should be identified and provided.

De-energisation is a process that is used to disconnect and isolate a system from a source of energy in order to prevent the release of that energy which may inadvertently, accidentally, or unintentionally cause harm to a person through movement or the release of heat, light, or sound.

In many countries, lockout tag-out (LOTO) is a legal requirement.

LOTO ensures that an equipment is not able to be started up before the completion of task(s) where maintenance or operations personnel may be exposed to the hazard of energised equipment or unintended energy releases.

A LOTO programme will help prevent:

- contact with a hazard while performing tasks that require the removal, bypassing, or deactivation of safeguarding devices
- unintended release of hazardous energy
- unintended start-up or motion of machinery, equipment, or processes

Components of a LOTO programme may include:

- LOTO process
- machine or task specific procedures
- LOTO training
- LOTO devices
- checking LOTO compliance

The inadvertent release of energy can result in serious accidents. Devices to control the release of energy should be provided.

A hazardous energy control programme is used to maintain worker safety by preventing:

- unintended release of stored energy
- unintended start-up
- unintended motion

- contact with a hazard when guards are removed or safety devices have been bypassed or removed

Because it is still possible for energy controls to fail, workers should be trained in avoiding the line of fire, where possible.

**Element Process Steps:**

1. identify hazardous energy sources
2. analyse work process and equipment
3. identify control measures
4. implement control measures
5. evaluate effectiveness and improve

**ISO 45001 Requirements:**

**6.1.2.1 Hazard identification**

The organisation shall establish, implement and maintain a process(es) for hazard identification that is ongoing and proactive. The process(es) shall take into account, but not be limited to:

f) other issues, including consideration of:

1) the design of work areas, processes, installations, **machinery/ equipment,** operating procedures and work organisation, including their adaptation to the needs and capabilities of the workers involved;

## ISO 45001 Annex A:

### A.6.1.2.1 Hazard identification

Hazards can be physical, chemical, biological, psychosocial, **mechanical, electrical** or based on movement and **energy**.

### E5. Asset Integrity Programme

*"Intact"*

### Element Description:

Asset integrity is the ability of an asset to perform its required function effectively and efficiently whilst safeguarding life and the environment.

An asset integrity programme should address the quality at every stage of the asset life cycle, from the design of new facilities to maintenance to decommissioning. Procedures to ensure the integrity of assets should be established.

An asset integrity programme usually includes inspections and audits.

Asset integrity programmes typically include:

- static equipment integrity
- rotating equipment integrity
- electrical equipment integrity
- instrument and control equipment integrity
- structural integrity
- mechanical integrity

Safety integrity levels (SIL) assessment should be carried out for safety instrumented system (SIS). SIL is a statistical representation of the availability of a required safety instrumented function (SIF).

A register of safety-related devices should be maintained.

**References:**

ISO 55001 Asset Management

IEC 61508 Functional Safety

**Element Process Steps:**

1. asset management policy
2. asset management strategy
3. asset management objectives
4. asset management plan
5. evaluate effectiveness and improve

**ISO 45001 Requirements:**

Nil.

**ISO 45001 Annex A:**

**A.8.1 Operational planning and control**

**A.8.1.1 General**

Operational planning and control of the processes need to be established and implemented as necessary to enhance occupational health and safety, by eliminating hazards or, if not practicable, by reducing the OH&S risks to levels as low as reasonably practicable for operational areas and activities.

Examples of operational control of the processes include:

c) establishing preventive or predictive maintenance and inspection programmes;

## E6  Control System for Safety

*"Control System"*

**Element Description:**

Equipment and processes need to be properly controlled to prevent accidents.

New equipment and facilities should be properly commissioned and checked that they are designed, installed, tested, operated, and maintained in accordance to requirements and specifications.

Control systems should comply with international standards such as IEC 61508 ("Functional safety of electric/electronic/programmable electronic safety-related systems"), IEC 61511, et cetera.

Control system (e.g., safety instrumented system) is also required to ensure a critical process can be put into a safe state in the event of operational problems upon demand. A safe state is a process condition where a hazardous HSE event cannot occur. A SIS is engineered to perform specific control functions to failsafe or maintain safe operation of a process when unacceptable or dangerous conditions occur.

Functional requirements and safety integrity requirements will need to be determined from hazard analysis. The safety integrity requirements may be verified by reliability analysis.

If cybersecurity is an issue and can affect the safety of an operations, standards such as IEC 62443 (Industrial communication networks—Network and system security) will be relevant. IEC 62443-2-1 defines the elements necessary to establish a cybersecurity management system (CSMS) for industrial automation and control systems (IACS) and provides guidance on how to develop those elements.

**Reference:**

IEC 60880 Nuclear power plants—Instrumentation and control systems important to safety—Software aspects for computer-based systems performing category A functions

IEC 61508 "Functional safety of electric/electronic/programmable electronic safety-related systems"

IEC 61511 "Functional safety—Safety instrumented systems for the process industry sector"

IEC 62443 "Industrial communication networks—Network and system security"

**Element Process Steps:**

1. identify relevant standards for managing control systems
2. hazard and risk assessment
3. allocate safety functions to protection layers
4. evaluate compliance with standards

5. conduct assessment of equipment

6. take actions to maintain compliance

7. evaluate effectiveness and improve

**ISO 45001 Requirements:**

**8.1.1 General**

The organisation shall plan, implement, control and maintain the processes needed to meet requirements of the OH&S management system, and to implement the actions determined in Clause 6, by:

a) establishing criteria for the processes;

b) implementing **control** of the processes in accordance with the criteria;

**8.1.2 Eliminating hazards and reducing OH&S risks**

The organisation shall establish, implement and maintain a process(es) for the elimination of hazards and reduction of OH&S risks using the following hierarchy of controls:

a) eliminate the hazard;

b) substitute with less hazardous processes, operations, materials or equipment;

c) use engineering **controls** and reorganisation of work;

### A.6.1.4 Planning action

When the assessment of OH&S risks and other risks has identified the need for controls, the planning activity determines how these are implemented in operation (see Clause 8); for example, determining whether to incorporate these **controls** into work instructions or into actions to improve competence.

### A.8.1 Operational planning and control

### A.8.1.1 General

Operational planning and control of the processes need to be established and implemented as necessary to enhance occupational health and safety, by eliminating hazards or, if not practicable, by reducing the OH&S risks to levels as low as reasonably practicable for operational areas and activities.

Examples of operational control of the processes include:

a) the use of procedures and systems of work;

### E7. Procurement Safety Programme

*"Buying safely"*

### Element Description:

Things (e.g., equipment, chemicals, supplies, raw materials, and other goods) that are being brought into a workplace may introduce safety and health hazards into an organisation.

An evaluation of the risk associated with planned purchases should be carried out. This screening may be done to determine the level of safety and health assessment that is needed. Items with significant OH&S risk should require an identification of OH&S hazards and related controls.

Procurement specification should include OH&S requirements such as:

- explosion-proof Class 1 Division 1
- DIN, EN, ISO standards, etc.

Specifications should be documented and maintained.

Appropriate risk-assessment methodology should be used for the assessment, which may include:

- failure mode and effects analysis
- HAZOP
- review of past related incidents

Legal and other requirements pertaining to the purchase should be identified by appropriate person. Necessary permits or licenses should be obtained where relevant.

Purchases should be checked for:

- conformance to specifications
- work as intended and function as designed
- usage requirements, precautions, or other protective measures are provided
- relevant safety information included

New chemicals and equipment to be used should be evaluated by an MOC process.

**Element Process Steps:**

1. establish procedure for procurement safety
2. identify products and services that require safety specifications
3. inspection and monitoring
4. evaluate effectiveness and improve

**ISO 45001 Requirements:**

**8.1.4 Procurement**

**8.1.4.1 General**

The organisation shall establish, implement and maintain a process(es) to control the procurement of products and services in order to ensure their conformity to its OH&S management system.

**ISO 45001 Annex A:**

**A.8.1 Operational planning and control**

**A.8.1.1 General**

Operational planning and control of the processes need to be established and implemented as necessary to enhance occupational health and safety, by eliminating hazards or, if not practicable, by reducing the OH&S risks to levels as low as reasonably practicable for operational areas and activities.

Examples of operational control of the processes include:

d) specifications for the **procurement** of goods and services;

### A.8.1.4 Procurement

### A.8.1.4.1 General

The **procurement** process(es) should be used to determine, assess and eliminate hazards, and to reduce OH&S risks associated with, for example, products, hazardous materials or substances, raw materials, equipment, or services before their introduction into the workplace.

The organisation's **procurement** process(es) should address requirements including, for example, supplies, equipment, raw materials, and other goods and related services purchased by the organisation to conform to the organisation's OH&S management system. The process should also address any needs for consultation (see 5.4) and communication (see 7.4).

The organisation should verify that equipment, installations and materials are safe for use by workers by ensuring:

a) equipment is delivered according to specification and is tested to ensure it works as intended;

b) installations are commissioned to ensure they function as designed;

c) materials are delivered according to their specifications;

d) any usage requirements, precautions or other protective measures are communicated and made available.

### F1. General Inspection Programme

*"Finding"*

### Element Description:

The main purpose of a general health and safety inspection programme is to identify and remove hazards and ensure safe working conditions.

Inspection may also be used to identify:

- unsafe behaviours,
- system deficiencies,
- noncompliance to requirements,
- poor housekeeping,
- cultural issues, and
- potential OH&S issues.

Specific checklists should be used for different areas if necessary if there are different characteristics.

Inspection findings should be properly tracked for closure and reviewed for effectiveness.

Critical areas should be inspected at least on a monthly basis.

**Element Process Steps:**

1. identify all areas requiring inspections
2. establish inspection programme
3. develop area specific inspection checklists
4. provide inspection training
5. implement programme

6. evaluate effectiveness and improve

**ISO 45001 Requirements:**

Nil.

**ISO 45001 Annex A:**

**A.6.1 Actions to address risks and opportunities**

**A.6.1.1 General**

OH&S opportunities address the identification of hazards, how they are communicated, and the analysis and mitigation of known hazards. Other opportunities address system improvement strategies.

Examples of opportunities to improve OH&S performance:

a) inspection and auditing functions;

**A.8.1.1 General**

Examples of operational control of the processes include:

c) establishing preventive or predictive maintenance and **inspection** programmes;

**F2. Signs, Colour-Coding, Labelling, and Tagging System**

*"See clearly"*

**Element Description:**

Safety signs warn workers who may be exposed to hazards in the workplace and communicate important instructions and PPE requirements.

Safety signs include:

- mandatory PPE signs (e.g., use helmet)
- safety equipment location signs (e.g., emergency shower, first aid station)
- prohibition signs (e.g., no smoking)
- danger signs (e.g., flammable materials, high voltage)
- warning signs (e.g., beware of opening door)
- notices (e.g., no entry)
- globally harmonised system (GHS) sign (e.g., corrosive, toxic)
- road signs (e.g., speed limit, no entry)

A colour-coding system quickly identify and draw attention to signs or hazards or the purpose of an object or equipment.

Colour coding may be used for:

- identifying hazards
- types of signs
- equipment (e.g., gas cylinders)
- status (e.g., green OK)
- inspection status of equipment (e.g., lifting gear)

- tools (e.g., explosive powered tool cartridge)
- pipelines (e.g., flammable fluids, fire quenching fluids)
- chemicals (e.g., flammable, toxic)
- PPE (e.g., helmets)
- electrical (e.g., wiring, voltage)
- floor markings (e.g., aisle, production, keep clear)

Labelling is critical to safety because it informs everyone as to the purpose of an object.

Labelling may be used for:

- chemicals (e.g., chemical name)
- materials (e.g. identification)
- containers (e.g., tanks, drums)
- storage (e.g., segregation)
- equipment (e.g., type)
- service (e.g., firefighting)
- purpose (e.g., pedestrian walkways)

A tagging system is important so that the correct equipment can be identified.

Equipment that is typically tagged:

- valves
- equipment (e.g., pumps, pressure vessels)

- instrumentation
- pipelines
- electrical equipment
- lifting gear

Equipment tagging system should be documented and communicated.

**Element Process Steps:**

1. identify needs for:
    - safety signs
    - colour coding
    - labelling and tagging
2. establish system
3. implement system
4. inspect and maintain
5. evaluate effectiveness and improve

**ISO 45001 Requirements:**

Nil.

## F3. Housekeeping, Order, and Cleanliness Programme

"5S"

### Element Description:

Order and good housekeeping is necessary for a safe and healthy work environment.

A typical housekeeping programme usually adopts the 5S methodology:

- sort (Seiri)
- set in order (Seiton)
- shine (Seiso)
- standard (Seiketsu)
- sustain (Shisuke)

Housekeeping programmes may include:

- worker training
- responsibilities
- housekeeping standards
- housekeeping schedule
- housekeeping inspections

### Element Process Steps:

1. identify all areas and activities requiring 5S

2. establish 5S requirements and programme

3. implement programme

4. conduct 5S inspections

5. evaluate effectiveness and improve

**ISO 45001 Requirements:**

Nil.

## F4. Management of Change and Pre-Startup Safety Review

*"Manage change"*

**Element Description:**

The purpose of the MOC programme is to prevent changes from inadvertently introducing hazards or risk into the operations. A MOC programme includes a review and authorisation process for evaluating changes to facility, operations, organisation, and more before implementation to ensure that unforeseen hazards are eliminated or controlled.

MOC programmes typically cover changes to:

- facility design

- operations

- organisation

The purpose of a PSSR is to ensure that any significant changes that are made to a facility or process equipment meet the original design or operating intent and that the process or facility is not required to operate in an unsafe condition. PSSR is a comprehensive review of new or modified facilities. The PSSR should confirm that:

- construction and equipment is in accordance with design specifications
- safety, operating, maintenance, and emergency procedures are in place and are adequate
- hazard analysis is conducted
- required actions have been closed
- MOC pre-start recommendations have been closed
- training has been completed

PSSR also serves as a quality check on the MOC completed.

**Element Process Steps:**

1. establish MOC and PSSR procedure(s)
2. provide training
3. implement procedure
4. monitor compliance
5. evaluate effectiveness and improve

**ISO 45001 Requirements:**

**6.1.2.1 Hazard identification**

The organisation shall establish, implement and maintain a process(es) for hazard identification that is ongoing and proactive. The process(es) shall take into account, but not be limited to:

g) actual or proposed changes in organisation, operations, processes, activities and the OH&S management system (see 8.1.3);

## 8.1.3 Management of change

The organisation shall establish a process(es) for the implementation and control of planned temporary and permanent changes that impact OH&S performance, including:

a) new products, services and processes, or changes to existing products, services and processes, including:

- workplace locations and surroundings;

- work organisation;

- working conditions;

- equipment;

- work force;

The organisation shall review the consequences of unintended changes, taking action to mitigate any adverse effects, as necessary.

## ISO 45001 Annex A:

### A.6.1.4 Planning action

Actions to address risks and opportunities should also be considered under the management of change (see 8.1.3) to ensure there are no resulting unintended consequences.

### A.8.1.3 Management of change

The objective of a management of change process is to enhance occupational health and safety at work, by minimizing the introduction of new hazards and OH&S risks into the work environment as changes occur (e.g. with technology, equipment, facilities, work practises and procedures, design specifications, raw materials, staffing, standards or regulations). Depending on the nature of an expected change, the organisation can use an appropriate methodology(ies) (e.g. design review) for assessing the OH&S risks and the OH&S opportunities of the change. The need to manage change can be an outcome of planning (see 6.1.4).

### F5. Safety Information System

*"Information"*

**Element Description:**

Effective management of OH&S requires information to be available for review and analysis. Safety information is required to help workers identify and understand the hazards from their work processes.

Safety information should include:

- information on the hazards of hazardous chemicals
- information on the technology of the process
- consequences of deviations
- information on the equipment and machines used

Safety information should be accessible to relevant personnel.

Safety information may include:

- safety data sheets
- equipment specifications
- design drawings

For ease of communication, certain safety information should be made available via the Internet or an intranet.

**Element Process Steps:**

1. identify all safety information
2. establish procedure to control all safety information
3. maintain safety information
4. evaluate effectiveness and improve

**ISO 45001 Requirements:**

**6.1.2.1 Hazard identification**

The organisation shall establish, implement and maintain a process(es) for hazard identification that is ongoing and proactive. The process(es) shall take into account, but not be limited to:

h) changes in knowledge of, and information about, hazards.

**7.5 Documented information**

**7.5.1 General**

The organisation's OH&S management system shall include:

b) documented information determined by the organisation as being necessary for the effectiveness of the OH&S management system.

### 7.5.3 Control of documented information

Documented information required by the OH&S management system and by this document shall be controlled to ensure:

Documented information of external origin determined by the organisation to be necessary for the planning and operation of the OH&S management system shall be identified, as appropriate, and controlled.

### ISO 45001 Annex A:

### A.6.1.2 Hazard identification and assessment of risks and opportunities

### A.6.1.2.1 Hazard identification

The organisation's hazard identification process(es) should consider:

f) changes in knowledge of, and information about, hazards:

1) sources of knowledge, information and new understanding about hazards can include published literature, research and development, feedback from workers, and review of the organisation's own operational experience;

2) these sources can provide new information about the hazards and OH&S risks.

## F6. Regulations, Codes and Internal Standards

*"Must do"*

### Element Description:

Applicable OH&S regulations, codes and standards should be identified and communicated for compliance by the organisation. A register of legal and other requirements should be established.

Organisations should identify or establish safety standards to be followed. Relevant guidelines and references which may provide guidance should also be identified.

Key areas of operations should be covered by relevant safety standards or guidelines.

Internal standards document minimum requirements for the organisation. Internal standards or references to external standards should be established for compliance.

Internal standards should comply with relevant requirements and industry regulations but they can be developed to be more specific to the organisation's needs and facilities.

Recognised and Generally Accepted Good Engineering Practises (RAGAGEP) applicable to the organisation should be identified.

### Element Process Steps:

1. establish procedure to identify all legal, other and internal requirements

2. communicate requirements to relevant personnel

3. evaluate compliance to requirements

4. Evaluate Effectiveness and Improve

**ISO 45001 Requirements:**

Nil

**ISO 45001 Annex A:**

**A.8.1 Operational planning and control**

**A.8.1.1 General**

Operational planning and control of the processes need to be established and implemented as necessary to enhance occupational health and safety, by eliminating hazards or, if not practicable, by reducing the OH&S risks to levels as low as reasonably practicable for operational areas and activities.

Examples of operational control of the processes include:

a) the use of procedures and systems of work;

b) ensuring the competence of workers;

c) establishing preventive or predictive maintenance and inspection programmes;

d) specifications for the procurement of goods and services;

e) application of **legal requirements and other requirements**, or manufacturers' instructions for equipment;

### A.9.1 Monitoring, measurement, analysis and performance evaluation

### A.9.1.1 General

In order to achieve the intended outcomes of the OH&S management system, the processes should be monitored, measured and analysed.

c) Examples of what could be monitored and measured to evaluate the fulfilment of other requirements can include, but are not limited to:

1) collective agreements (when not legally binding);

2) **standards and codes;**

d) Criteria are what the organisation can use to compare its performance against.

1) Examples are benchmarks against:

i) other organisations;

ii) **standards and codes;**

iii) **the organisation's own codes and objectives;**

F7. **Capital Projects Safety Management**

*"Safe projects"*

**Element Description:**

Capital expenditure projects will introduce a spectrum of OH&S risks into an organisation. If safety can be considered and managed at the earliest stages of a project, many risks may be eliminated at the design stage and less OH&S issues will be encountered at later stages.

Project management should consider safety at each stage of a project with actions taken to eliminate or reduce the risk.

"Design for safety" should be considered at relevant stages of the project.

**Element Process Steps:**

1. establish procedure to manage projects
2. establish specific safety requirements for each project
3. establish project risk register
4. evaluate compliance with all legal and other requirements upon completion
5. evaluate effectiveness and improve

**ISO 45001 Requirements:**

Nil.

**ISO 45001 Annex A:**

**A.6.1 Actions to address risks and opportunities**

**A.6.1.1 General**

Examples of other opportunities to improve OH&S performance:

— integrating occupational health and safety requirements at the earliest stage in the life cycle of facilities, equipment or process planning for facilities relocation, process re-design or replacement of machinery and plant;

— integrating occupational health and safety requirements at the earliest stage of planning for facilities relocation, process re-design or replacement of machinery and plant;

**A.6.1.2 Hazard identification and assessment of risks and opportunities**

**A.6.1.2.1 Hazard identification**

The ongoing proactive identification of hazard begins at the conceptual design stage of any new workplace, facility, product or organisation. It should continue as the design is detailed and then comes into operation, as well as being ongoing during its full life cycle to reflect current, changing and future activities.

**A.8.1.4 Procurement**

**A.8.1.4.1 General**

The procurement process(es) should be used to determine, assess and eliminate hazards, and to reduce OH&S risks associated with, for example, products, hazardous materials or substances, raw materials, equipment, or services before their introduction into the workplace.

The organisation's procurement process(es) should address requirements including, for example, supplies, equipment, raw materials, and other goods and related services purchased by the organisation to conform to the organisation's OH&S management system. The process should also address any needs for consultation (see 5.4) and communication (see 7.4).

The organisation should verify that equipment, installations and materials are safe for use by workers by ensuring:

a) equipment is delivered according to specification and is tested to ensure it works as intended;

b) installations are commissioned to ensure they function as designed;

c) materials are delivered according to their specifications;

d) any usage requirements, precautions or other protective measures are communicated and made available.

## G - GENERAL

### G1. Operating Procedures, Instructions, and Method Statements

*"Steps"*

**Element Description:**

Operating procedures are written instructions that document the steps for a task and describe how the steps are to be carried out.

Well-written procedures will also include the safety precautions to be taken and required personal protective equipment. Procedures may also provide information on troubleshooting any deviations or anomalies. Procedures provide assurance that workers will execute a task in a consistent and intended manner.

Procedures should be:

- properly developed and controlled
- communicated
- included in training

- periodically reviewed

Some work need not be proceduralised if the sequence is not important and simple practices, rules, or guidelines are adequate. Rules and practices can also cover a wider range of activities. The purpose of safe work practices is to control risk associated with nonroutine work or where the work may vary due to different circumstances. Routine activities are usually covered by operating procedures if necessary.

Examples:

- lockout tag-out
- confined space entry
- access to process area
- excavation in process area
- lifting operations

Work procedures may have different names and varying formats. They may be called:

- work instructions
- operating procedure
- method statement
- safe work procedure

However, these terms may also have slightly different meanings in different organisations.

Procedures may be developed together with job hazard analysis (JHA).

## Element Process Steps:

1. identify all activities requiring a procedure
2. establish a format for procedures
3. identify procedures that need document control
4. review and approve procedures
5. evaluate effectiveness and improve

## ISO 45001 Requirements:

### 8.1 Operational planning and control

### 8.1.1 General

The organisation shall plan, implement, control and maintain the processes needed to meet requirements of the OH&S management system, and to implement the actions determined in Clause 6, by:

a) establishing criteria for the processes;

b) implementing control of the processes in accordance with the criteria;

c) maintaining and retaining documented information to the extent necessary to have confidence that the processes have been carried out as planned;

**ISO 45001 Annex A:**

**A.8.1 Operational planning and control**

**A.8.1.1 General**

Operational planning and control of the processes need to be established and implemented as necessary to enhance occupational health and safety, by eliminating hazards or, if not practicable, by reducing the OH&S risks to levels as low as reasonably practicable for operational areas and activities.

Examples of operational control of the processes include:

a) the use of procedures and systems of work;

## G2. Emergency Preparedness and Crisis Management

*"Ready"*

**Element Description:**

Appropriate response to an emergency can save lives and minimise property damage and losses.

During an emergency, there is insufficient time for planning, discussions, and training. Emergency preparedness puts an organisation into a state of readiness to respond to an emergency.

An emergency preparedness programme typically includes:

- system identification of emergency scenarios
- organisation for emergency response
- formation of emergency response team(s)
- emergency plans

- training of emergency team(s)
- review of emergency response needs and equipment
- inspection and maintenance of emergency equipment
- emergency drills, tests, exercises, and post-mortem
- mutual aid arrangements
- coordination with external responders

Crisis management is the application of strategies designed to help an organisation deal with a sudden and significant event. An emergency may escalate into a crisis. A crisis, if poorly managed, may threaten the organisation or its stakeholders. Crisis management is an escalation of emergency response and required the sustained involvement of senior management until the episode is over.

Elements of crisis management over emergency response are:

- crisis management team
- crisis management plans
- communication with stakeholders
- coordination with external agencies
- handling of mass media and social media
- external communication
- communication with next of kin

Business continuity is the planning and preparation of an organisation to make sure that it overcomes serious incidents, disasters, or disruptions so that it resumes its normal operations in as short a period as possible, regardless of the cause of the emergency.

A business continuity management (BCM) process in an organisation will strengthen the overall response to emergencies and incidents.

A BCM typically includes the following:

- business impact analysis
- business continuity options
- exercise and testing
- recovery time objective
- minimum business continuity objective
- minimum acceptable level of operations
- minimum operating requirements
- alternative arrangements
- contractual arrangements

**Reference:**

ISO 22301 Business Continuity Management

**Element Process Steps:**

1. identify all potential emergency situations
2. establish specific emergency plans
3. prepare equipment and train personnel

4. test emergency plans

5. conduct periodic drills and post-mortem

6. evaluate effectiveness and improve

**ISO 45001 Requirements:**

**6.1.2.1 Hazard identification**

The organisation shall establish, implement and maintain a process(es) for hazard identification that is ongoing and proactive. The process(es) shall take into account, but not be limited to:

d) potential emergency situations;

**6.1.4 Planning action**

The organisation shall plan:

a) actions to:

3) prepare for and respond to emergency situations (see 8.2);

**8.2 Emergency preparedness and response**

The organisation shall establish, implement and maintain a process(es) needed to prepare for and respond to potential emergency situations, as identified in 6.1.2.1, including:

a) establishing a planned response to emergency situations, including the provision of first aid;

b) providing training for the planned response;

c) periodically testing and exercising the planned response capability;

d) evaluating performance and, as necessary, revising the planned response, including after testing and, in particular, after the occurrence of emergency situations;

e) communicating and providing relevant information to all workers on their duties and responsibilities;

f) communicating relevant information to contractors, visitors, emergency response services, government authorities and, as appropriate, the local community;

g) taking into account the needs and capabilities of all relevant interested parties and ensuring their involvement, as appropriate, in the development of the planned response.

**ISO 45001 Annex A:**

**A.6.1.2 Hazard identification and assessment of risks and opportunities**

**A.6.1.2.1 Hazard identification**

The organisation's hazard identification process(es) should consider:

d) potential emergency situations:

1) unplanned or unscheduled situations that require an immediate response (e.g. a machine catching fire in the workplace, or a natural disaster in the vicinity of the workplace or at another location where workers are performing work-related activities);

2) include situations such as civil unrest at a location at which workers are performing workrelated activities which requires their urgent evacuation;

### A.6.1.4 Planning action

The actions planned should primarily be managed through the OH&S management system and should involve integration with other business processes, such as those established for the management of the environment, quality, **business continuity, risk,** financial or human resources. The implementation of the actions taken is expected to achieve the intended outcomes of the OH&S management system.

### A.8.2 Emergency preparedness and response

Emergency preparedness plans can include natural, technical and man-made events that occur inside and outside normal working hours.

### A.9.1 Monitoring, measurement, analysis and performance evaluation

### A.9.1.1 General

In order to achieve the intended outcomes of the OH&S management system, the processes should be monitored, measured and analysed.

a) Examples of what could be monitored and measured can include, but are not limited to:

3) the effectiveness of operational controls and **emergency exercises**, or the need to modify or introduce new controls;

## G3 Permit to Work System

*"Authorisation"*

### Element Description:

The purpose of a permit to work (PTW) system is to control specific types of work that are potentially hazardous. Basically, in a PTW system, permit or authorisation is required before the specified work may proceed. Authorisation is only given after a safety assessment is carried out to verify that conditions are safe for work and all safety measures have been complied with.

PTW systems typically cover:

- confined space entry
- lifting operations
- cold work
- hot work
- excavation
- work at height
- electrical supply work
- bypass of safety interlock
- vehicle entry to process area
- radiography work

A PTW system may be electronic.

### Element Process Steps:

1. identify the needs for permits to work
2. establish PTW procedure
3. implement PTW procedure
4. audit PTW compliance
5. evaluate effectiveness and improve

### ISO 45001 Requirements:

Nil.

### ISO 45001 Annex A:

**A.6.1 Actions to address risks and opportunities**

**A.6.1.1 General**

OH&S opportunities address the identification of hazards, how they are communicated, and the analysis and mitigation of known hazards. Other opportunities address system improvement strategies.

Examples of opportunities to improve OH&S performance:

d) **permit to work** and other recognition and control methods;

## G4. Contractor Management Programme

*"Working safely together"*

**Element Description:**

Contractors working on-site may be endangered by the hazards present, or they may cause an accident and endanger others.

There is a wide variation in how a contractor may work in an organisation:

- working independently or with client's workers
- ad-hoc or routine work
- low-risk or high-risk
- skilled labour or unskilled work
- self-directed or supervised
- proportion of work by subcontractors
- level of safety supervision
- own equipment or client's equipment
- paid by rate or lump sum
- long or short duration
- single or multinationality workforce
- familiar or unfamiliar work environment
- need for coordination and PTWs

The system to manage contractors should take into account the above factors.

A contractor management programme includes:

- selection of contractors
- evaluation of contractors
- control of contractors
- competence of contractors' personnel
- safety induction of contractors' personnel

**Element Process Steps:**

1. establish a contractor management procedure to cover:
   - selection of contractor
   - control of contractor
   - performance evaluation of contractor
2. evaluate, approve, and register contractors
3. monitor and evaluate contractors' safety performance
4. evaluate effectiveness and improve

**ISO 45001 Requirements:**

**4.3 Determining the scope of the OH&S management system**

The OH&S management system shall include the activities, products and services within the organisation's control or influence that can impact the organisation's OH&S performance.

### 8.1.4.2 Contractors

The organisation shall coordinate its procurement process(es) with its contractors, in order to identify hazards and to assess and control the OH&S risks arising from:

a) the contractors' activities and operations that impact the organisation;

b) the organisation's activities and operations that impact the contractors' workers;

c) the contractors' activities and operations that impact other interested parties in the workplace.

The organisation shall ensure that the requirements of its OH&S management system are met by contractors and their workers. The organisation's procurement process(es) shall define and apply occupational health and safety criteria for the selection of contractors.

### ISO 45001 Annex A:

### A.8.1.4.2 Contractors

Assignment of activities to contractors does not eliminate the organisation's responsibility for the occupational health and safety of workers.

An organisation can achieve coordination of its contractors' activities through the use of contracts that clearly define the responsibilities of the parties involved. An organisation can use a variety of tools for ensuring contractors' OH&S performance in the workplace (e.g. contract award mechanisms or prequalification criteria which consider past health and safety performance, safety training, or health and safety capabilities, as well as direct contract requirements).

When coordinating with contractors, the organisation should give consideration to the reporting of hazards between itself and its contractors, controlling worker access to hazardous areas, and procedures to follow in emergencies. The organisation should specify how the contractor will coordinate its activities with the organisation's own OH&S management system processes (e.g. those used for controlling entry, for confined space entry, exposure assessment and process safety management) and for the reporting of incidents.

The organisation should verify that contractors are capable of performing their tasks before being allowed to proceed with their work; for example, by verifying that:

a) OH&S performance records are satisfactory;

b) qualification, experience and competence criteria for workers are specified and have been met (e.g. through training);

c) resources, equipment and work preparations are adequate and ready for the work to proceed.

## G5. Investigations and Learning from Incident System

*"No recurrence"*

### Element Description:

A key part of safety management is to learn from incidents (LFI) that have occurred within and outside the organisation to prevent similar accidents from happening.

Important lessons can only be learnt if it is supported by good reporting from workers and a robust investigation process.

The investigation process must be able to identify all the multiple causes that may contribute to an incident. Selected members of the investigation team should be trained in incident investigation techniques.

Incident investigation should be supported by an appropriate root cause analysis methodology (Element A5).

For such a system to be successful, there must be willingness to share the information relating to incidents that have occurred.

The following information should be available:

- nature of incident
- context of the incident
- causes of the incident
- possible actions to prevent similar incident

All incidents and lessons learnt should be communicated to all relevant persons.

Information from incident investigation is communicated in LFI. External incidents will rely on the investigations by external parties. Sources or databases of useful information of external incidents should be identified and checked periodically.

Sufficient information should be provided so that preventive actions can be taken.

LFI system provides opportunities for improvement and is also important because it is a reminder of the possibility of accident and everyone should remain constantly alert.

**Element Process Steps:**

1. establish incident investigation procedure

2. train relevant personnel in incident investigation and root cause analysis

3. promote incident reporting

4. communicate "lessons learnt" and LFI

5. evaluate effectiveness and improve

**ISO 45001 Requirements:**

**6.1.2.1 Hazard identification**

The organisation shall establish, implement and maintain a process(es) for hazard identification that is ongoing and proactive. The process(es) shall take into account, but not be limited to:

c) past relevant incidents, internal or external to the organisation, including emergencies, and their causes;

**7.3 Awareness**

Workers shall be made aware of:

d) incidents and the outcomes of investigations that are relevant to them;

## G6. Leading Indicators and Monitoring

*"Forecast"*

**Element Description:**

Lagging indicators such as accident frequency rates are reactive and "after the fact" and do not help in preventing accidents. Leading indicators are proactive because they provide an indication on the status of the OH&S management system in terms of performance and effectiveness in preventing accidents.

Leading indicators are measures of safety activities and safety effort preceding any incidents.

Focus on leading indicators avoids knee-jerk, "after the fact" responses to accidents and the neglect of safety effort when there is no accident.

Typical examples of leading indicators include:

- training plan completion
- inspection programme compliance
- quality of Inspections
- number of safety improvement projects
- JHA completion (percentage)
- emergency drills (percentage)
- number of safety promotional activities
- maintenance programme compliance (percentage)
- number of BBS observations

**Element Process Steps:**

1. identify leading indicators
2. establish reporting system for leading indicators
3. collect data

4. evaluate effectiveness and improve

**ISO 45001 Requirements:**

**9.1 Monitoring, measurement, analysis and performance evaluation**

**9.1.1 General**

The organisation shall establish, implement and maintain a process(es) for monitoring, measurement, analysis and performance evaluation.

The organisation shall determine:

a) what needs to be monitored and measured, including:

- 1) the extent to which legal requirements and other requirements are fulfilled;

- 2) its activities and operations related to identified hazards, risks and opportunities;

- 3) progress towards achievement of the organisation's OH&S objectives;

- 4) effectiveness of operational and other controls;

b) the methods for monitoring, measurement, analysis and performance evaluation, as applicable, to ensure valid results;

c) the criteria against which the organisation will evaluate its OH&S performance;

d) when the monitoring and measuring shall be performed;

e) when the results from monitoring and measurement shall be analysed, evaluated and communicated.

The organisation shall evaluate the OH&S performance and determine the effectiveness of the OH&S management system.

### ISO 45001 Annex A:

**A.6.2.1 OH&S objectives**

Objectives are established to maintain and improve OH&S performance. The objectives should be linked to risks and opportunities and **performance criteria** which the organisation has identified as being necessary for the achievement of the intended outcomes of the OH&S management system.

**A.9.1 Monitoring, measurement, analysis and performance evaluation**

**A.9.1.1 General**

In order to achieve the intended outcomes of the OH&S management system, the processes should be monitored, measured and analysed.

d) Criteria are what the organisation can use to compare its performance against.

1) Examples are benchmarks against:

iv) OH&S statistics.

2) To measure criteria, indicators are typically used; for example:

ii) if the criterion is a comparison of completions of corrective actions, then the **indicator could be the percentage** completed on time.

Measurement generally involves the assignment of numbers to objects or events. It is the basis for quantitative data and is generally associated with the **performance evaluation of safety programmes** and health surveillance.

**Performance evaluation** is an activity undertaken to determine the suitability, adequacy and effectiveness of the subject matter to achieve the established objectives of the OH&S management system.

## G7. New Technology, Improvement, and Benchmarking

*"Good, better, best"*

### Element Description:

The world undergoes constant technological change and safety problems that cannot be solved yesterday may be solvable tomorrow. Solutions that were prohibitively expensive previously may become affordable.

Periodic reviews of risk control measures should be carried out as more effective and efficient methods may become suddenly available.

Company should have processes that continually identify new areas for safety improvement and adoption of new technology and methods.

Safety improvement projects and teams provide a focus for continual improvement efforts. OH&S issues needing improvements should be identified by processes and feedback.

Improvement teams should be appointed to study critical OH&S problems.

Various improvement tools, techniques, and approaches may be used:

- kaizen
- error proofing
- six sigma
- flowcharts
- fishbone diagram
- charts and diagrams

Technological solutions mentioned earlier included:

- Internet of things (IoT) and Internet of everything (IoE)
- cameras and CCTV
- instantaneous communication
- data transmission and networking,
- Internet
- RFID, QR, and barcodes
- handheld devices
- biometric recognition
- GPS tracking devices
- software programmes for MOC, PTW, etc.
- safety management software

- drones

**Element Process Steps:**

1. form safety improvement teams for critical OH&S issues

2. assign personnel to monitor
    - technological development
    - industry standards
    - international best practices

3. appoint representative to participate in industry groups

4. evaluate effectiveness and improve

**ISO 45001 Requirements:**

**5.1 Leadership and commitment**

Top management shall demonstrate leadership and commitment with respect to the OH&S management system by:

h) ensuring and promoting continual improvement;

**7.4.2 Internal communication**

The organisation shall:

b) ensure its communication process(es) enables workers to contribute to continual improvement.

### 6.1.2.3 Assessment of OH&S opportunities and other opportunities for the OH&S management system

The organisation shall establish, implement and maintain a process(es) to assess:

a) OH&S opportunities to enhance OH&S performance, while taking into account planned changes to the organisation, its policies, its processes or its activities and:

1) opportunities to adapt work, work organisation and work environment to workers;

2) opportunities to eliminate hazards and reduce OH&S risks;

b) other opportunities for improving the OH&S management system.

### 8.1.3 Management of change

The organisation shall establish a process(es) for the implementation and control of planned temporary and permanent changes that impact OH&S performance, including:

d) developments in knowledge and technology.

### 10 Improvement

### 10.1 General

The organisation shall determine opportunities for improvement (see Clause 9) and implement

necessary actions to achieve the intended outcomes of its OH&S management system.

## 10.3 Continual improvement

The organisation shall continually improve the suitability, adequacy and effectiveness of the OH&S management system, by:

a) enhancing OH&S performance;

### **ISO 45001 Annex A:**

### A.6.1 Actions to address risks and opportunities

### A.6.1.1 General

Examples of other opportunities to improve OH&S performance:

— using new **technologies** to improve OH&S performance;

— **benchmarking**, including consideration of both the organisation's own past performance and that of other organisations;

### A.9.1 Monitoring, measurement, analysis and performance evaluation

### A.9.1.1 General

In order to achieve the intended outcomes of the OH&S management system, the processes should be monitored, measured and analysed.

d) Criteria are what the organisation can use to compare its performance against.

1) Examples are **benchmarks** against:

      i) other organisations;

ii) standards and codes;

iii) the organisation's own codes and objectives;

iv) OH&S statistics.

## A.10 Improvement

### A.10.1 General

The organisation should consider the results from **analysis and evaluation** of OH&S performance, evaluation of compliance, internal audits and management review when taking action to improve.

www.ingramcontent.com/pod-product-compliance
Lightning Source LLC
Chambersburg PA
CBHW020732180526
45163CB00001B/207